Mansour Eslami

Senior Design Experience: Lessons for Life

Agile Press
A Division of *C* FAR
Center For Agile Research
Chicago, Illinois

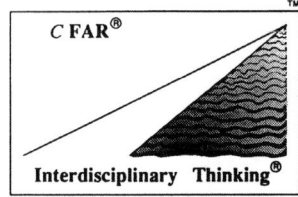

Center For Agile Research, Inc.
Corporate Headquarters
P.O. Box 1939, Chicago, IL 60690-1939

Professor Mansour Eslami
Dept. of Electrical and Computer Engineering
The University of Illinois at Chicago
Chicago, Illinois 60607
USA
eslami@ieee.org
January 24, 2002

ISBN 0-9718239-0-1 Agile Press, Chicago, Illinois, U.S.A.

Library of Congress Control Number: 2002090654

Cataloging-in-Publication Data:
Eslami, Mansour
Senior Design Experience: Lessons for Life
ISBN 0-9718239-0-1
1. Technical Writing. 2. Engineering Management.
3. Engineering Economics. 4. Intellectual Property and Patent.
5. Ethics. 6. Engineering Design. 7. Professional Liability for Engineers.
8. Marketing and Business Planning. 9. Entrepreneurial and Financing a Venture.
I. Title. II. Index. III. References.

This work is subject to copyright. All rights are reserved, whether the whole or part of the material is concerned, specifically the rights of translation, reprinting, reuse of illustrations, recitation, broadcasting, reproduction on microfilm or in other ways, and storage in data banks. Duplication of this publication or parts thereof is permitted only under the provisions of the United States Copyright law in its current version, and permission for use must always be obtained from the publisher.
    This textbook contains information obtained from original and highly respected sources. Reprinted materials are quoted with permission and consentingly as indicated. Every effort is made to publish reliable and accurate information, however, neither the author nor the publisher assume any responsibility for the validity of all materials or consequences of their use.

The use of general descriptive names, registered names, trademarks, *etc.,* in this publication does not imply, even in the absence of specific statement, that such names are exempt from the relevant protective laws and regulations and therefore free for general use. These are used only for identification and explanation, without intent to infringe.

Copyright © 2002 by Agile Press, a Division of *C* FAR, Center For Agile Research, Chicago, Illinois.
Please direct all inquiries to the address above, or *agilepress@agileresearch.com*

Printed in Chicago, Illinois, USA
10 9 8 7 6 5 4 3 2 1
Typesetting: Camera-ready by the author

Part Zero: Textbook Structure

Table of Contents

Preface, vi
Acknowledgment, vii

Part One: Course Administrative Issues

- *1.1. Introduction, 3*
- *1.2. ABET Requirements, 3*
- *1.3. General Guidelines, 4*
- *1.4. Typical Catalog Course Description, 5*
- *1.5. The Core of Senior Design Courses, 5*
- *1.6. Senior Design Contract, 7*
- *1.7. Steps to Select a Project, 9*

Part Two: Course Lecture Notes

Lecture Notes – A Summary, 13
Lecture One: Introduction and Course Outline, 15
Lecture Two: Preliminary Issues, Expectations and Professionalism, 21
Lecture Three: Final List of Projects and Contracts, 24
Lecture Four: Laboratory Notebook and Proposals, 27
Lecture Five: Proposals and Memos Continued, 34
Lecture Six: Steps Towards Oral Presentations – Phase I, 36
Lecture Seven: Technical Writing – Phase I, 39
Lecture Eight: Marketing High - Tech Products, 42
Lecture Nine: Marketing High - Tech Products (Continued), 50
Lecture Ten: In Preparation for the Design Review, 56
Lecture Eleven: Ethics, 58
Lecture Twelve: Design Review Preparation – The Final Thought, 60
Lecture Thirteen: Mid - Course Design Review Results, 63
Lecture Fourteen: Technical Writing – Phase II, 66
Lecture Fifteen: In Preparation for the Final Report, 72
Lecture Sixteen: Steps Towards Oral Presentation – Phase II, 79
Lecture Seventeen: Engineering Economics, 81
Lecture Eighteen: Entrepreneurial and Venture Capital, 89
Lecture Nineteen: Entrepreneurial and Venture Capital (Continued), 99
Lecture Twenty: Quality Control, 108
Lecture Twenty One: Intellectual Property Issues, 123
Lecture Twenty Two: Intellectual Property and Patents, 139
Lecture Twenty Three: Risk Management, 149
Lecture Twenty Four: In Preparation for the Final Presentation, 159

Part Three: Samples of Student's Work

Introduction, 163
Who, when, where and how did it start? 164

- I. **Mid-Course Report, 165**
  - [1] Naqeeb Ameeri and Ali Seraj, ***Motion Detection and Object Tracking in Real Time, 166***
- II. **Final Presentations Slides**, *175*
  - [1] Christopher Bertell, Hiep Nguyen, Kathryn Wassink, ***Machine Vision Interface with Seiko Robot Arm, 176***
  - [2] Kamal Roumani and Jeremiah Kent, ***Robust Control Testbed, 183***
  - [3] Naqeeb Ameeri and Ali Seraj, ***Motion Detection and Object Tracking in Real Time, 190***
  - [4] Karl Doering, Nadim Addus, Heath Deuel, Steven Gyford, Oscar Servin, Michael Wilson, ***Wireless Communication Systems Block Analysis and Design, 197***
- III. **A sample page of entries in the journal of Arun Gupta and Phillip Hoang**, *219*
- IV. **Final Report, 220**
  - [1] Karl Doering, Nadim Addus, Heath Deuel, Steven Gyford, Oscar Servin, Michael Wilson, ***Wireless Communication Systems Block Analysis and Design, 221***

Index, 279

Preface

Since the publication of my textbook, *Analog and Digital Circuits Theory and Experimentation,* in 1987, I was ambitiously hoping to reach every engineering student by writing a textbook that becomes the primary source of information for years to come. I wanted to complement the previous textbook such that every student benefits from and keeps that for life. The initial idea to write this textbook came to me when I learned that there are concerns among educators about two concepts: teaching new senior design courses; and meeting the requirements for the Accreditation Board for Engineering and Technology (ABET). It was not really clear where to look for information to develop such new courses, which meet the recent recommendations of this board (ABET). Thus, to address commonalties in these design courses became the main focus of my activity and the main theme of this textbook. These materials are gathered in response to the expectations and requirements of ABET. Although this textbook is not related to or endorsed by that organization, every effort is made to meet their criteria as best as one could have understood.

In this regard a concise account of various underlying issues that comprises this course is described in Part One. Issues pertaining to the formation of the corresponding lecture notes, are described in Part Two, followed by samples of student's work in Part Three. I believe students should know various constraints associated with this course in order to plan their work accordingly and that is why Part One is included. Although this book is written to address senior students primarily, it is my sincere hope that it will help instructors of similar courses in their planning. As such I have left ample opportunities for them to revise my presentations, based on their need and experience. I am looking forward to receiving their constructive suggestions for improvement of this textbook, which I will incorporate graciously in its future editions.

Thus, based on many years of teaching experience in engineering schools, I developed this textbook, which I hope fits the bill and addresses commonalties in these design courses.

Mansour Eslami
Dept. of Electrical and Computer Engineering
The University of Illinois at Chicago
Chicago, Illinois 60607
eslami@ece.uic.edu
January 24, 2002

Acknowledgement

It is my pleasure to thank all staff at the University of California at Riverside (UCR) who helped me with the creation of this series of design courses during Winter and Spring of 2001, when I was visiting the Dept. of Electrical Engineering. In particular, the initial web sites on WCB.UCR, for EE175A/B, was developed with substantial help and guidance of Dr. Leo Schouest who was constantly an inspiring colleague, and my former Teaching Assistant, Mr. Kirill Shabunov. Because of those web sites, I had to rewrite the notes into the current form to improve my earlier work. Thanks to all staff at UCR for their collective efforts to encourage me to complete this task, especially Ms. Zina Romero, Ms. Janet Harshman, and Mr. Isaac Saldana.

Most importantly, special thanks are due to my former students in EE175A/B at UCR. This book is dedicated to them for their hard work and sincerity. However, it would be out of character, if I would not thank all my former students worldwide – about 4500 strong so far. Thus, it is always a pleasure to thank them all for keeping me going all these years. They come from almost every country in the world. I have been the subject of many high-level family discussions worldwide, since many of them have a story about me to tell.

Special thanks are also due to my colleagues at the University of Illinois at Chicago for their very timely request that I finish this work and get this book ready for students in this and other universities around the country.

These days nothing can be done without extensive support of computer staff. I am grateful to my good friends currently at UIC, Ralph Orlick, Certa Nicholas, and Richard Chang. Also, I have not forgotten the help and support of my good friends and former staff at UIC, Leland Luecke, Ashish Desai, John Yancey and Clarence Murzyn, who had set up my other computer networks. I thank Barbara Sykes for her tremendous help in this project as well.

Finally, special thanks to my colleagues at *C* FAR, especially my dear friends Richard and Mildred Bleier, for supporting this project enthusiastically.

Mansour Eslami
Dept. of Electrical and Computer Engineering
The University of Illinois at Chicago
Chicago, Illinois 60607
eslami@ece.uic.edu
January 24, 2002

Part One: Course Administrative Issues

1.1. Introduction

In this part, a concise account of various underlying issues that comprises this course and the formation of the corresponding lecture notes are described. These materials are gathered in response to the expectations and requirements of accreditation board for engineering and technology (ABET). Although we are not representing, nor this textbook is related to or endorsed by that organization, we have made every effort to meet their criteria as best as we have understood.

This part opens with a brief overview of ABET requirements; some general guidelines, and a typical university catalog course description for senior design. Then follows by the core of senior design courses as we have understood, and a typical senior design contract that is described subsequently. Part one ends with a procedure for students on steps to select a project.

1.2. ABET Requirements

Very briefly, engineering students must have a sound training in mathematics (including probability and statistics), natural sciences, computer sciences, humanities, social sciences, communicational skills, and engineering topics including engineering science and engineering design. Engineering science is incorporated by creative application of basic science skills, and engineering design focuses on the process of devising a system, or component, or process. Design element is incorporated into the program by a team of faculty and concentrates on design issues of a real system through construction and possible fabrication of a working prototype. This design element is advanced through the use of mathematics and sciences as well as the use of design tools and analytical skills during a series of project courses, which will culminate in a Capstone senior design project. In short, students will continually be informed of the ultimate goal of being an engineer through a series of design courses and working in a team. This textbook is designed to elaborate and meet various requirements pertaining to this series of courses.

1.3. General Guidelines

Objectives. The Senior Design Project is the culmination of course work in the bachelor's degree program in each field of engineering. In this comprehensive two-semester or two-quarter course, students are expected to apply the concepts and theories of their discipline to a novel research project. A written report, giving details of the project and test results, and an oral presentation giving the details of the project, will be required to complete this course satisfactorily.

Credit and Hours. Up to five semester (or up to eight quarter) units of engineering design credit will be granted for the completed project. It is expected that up to twelve hours of laboratory (or field) work will be required weekly for satisfactory completion of the project. The design value of these units has been accounted for in the total number of required science and design units necessary for graduation to meet the ABET requirements.

Project Participants. Projects may be completed individually (though not recommended), or in small teams with shared responsibility. In the team option, each student will be held responsible for a distinct component of the total team effort. Team projects are sufficiently more complex than individual projects so as to allow for an appropriate work load for all team members.

Technical Faculty Supervisors. All projects will have a technical supervisor who will serve as a resource person and sponsor, however, the instructor of the course will be responsible for approving the proposal and assuring that the project is of sufficient content and work load to satisfy the course objectives. Individuals or teams are to secure sponsorship by the beginning of the first semester or the first quarter.

Project Topics. Engineering projects will include proposal, design, development of software/hardware, and testing of the corresponding engineering devices or systems. Possible project topics may be obtained from the engineering faculty, or from cooperating industry. However, students may also propose an original project of their own.

Weekly Class Meeting. The entire class of senior design will meet once each week for one hour. These meetings are intended to provide instruction in topics common to all design projects (engineering economics, ethics, etc.). They may include brief presentations by each team, aimed at improving technical presentation skills. The class coordinator will conduct the sessions. These meetings are mandatory and are for student's benefit.

1.4. Typical Catalog Course Description

Senior Design Project (four to five semester-hour credits or up to eight quarter-hour units) with nine hours Laboratory work plus one hour consultation per week. Under the direction of a faculty member, students (individually or in small teams with shared responsibilities) propose, design, build, and test engineering devices or systems. Major design experience for senior students through projects with local industries can also be arranged. Open ended design experience requiring the application of a broad spectrum of the student's engineering knowledge to a practical design problem. A written report, giving details of the project and test results, and an oral presentation of the design aspects are required. An in progress or deferred grade will be given for the first course initially. A letter grade will be given for both courses finally. Prerequisite: Senior standing in Engineering.

1.5. The Core of Senior Design Courses

The Senior Design Projects must be administratively organized and supervised by one faculty member who also grades the entire class, while other faculty members equitably contribute to the technical supervision of students. The Senior Design Project will include the project specification, design, budget, reporting, and presentation requirements specified by the supervising faculty member and the project contract shown subsequently in this part, which must also be signed accordingly. These courses will include a lecture component of at least one hour per week in order to meet ABET requirements. Topics to be covered include project management, budgeting, scheduling, engineering economics, and engineering ethics plus similar topics as the supervising faculty deems necessary; based on the particular strength and/or needs of the underlying engineering program.

A Student Group – This may include students from other engineering disciplines. All students in a group (no more than four persons per group), however, must have a well-defined role. On such occasions when students come from different engineering programs, each department will grade its own students.

Senior Design – This is a design course and students must define a *design* project, not a *research*, nor an *evaluation*, nor a *fabrication* project. It is rather a balanced approach to culminate in using the many of these ideas.

Senior Design Projects will be from one of the following categories:

(1) Faculty Specified Projects – Students can work on projects with the department faculty members to be known as a technical advisor. This requires a signed agreement by that faculty, the student(s), and the overall course supervising faculty member. The technical advisor is expected to meet with the group weekly to provide technical leadership. Faculty members outside of the department or even outside of the College of Engineering can also supervise senior design students, provided that the student group also arranges for setting up technical meetings within the College of Engineering faculty, since that may become necessary to complete the design project rigorously. Students on such projects will still be required to attend lecture and meet weekly with the supervising faculty member to report on their progress. The supervising faculty will be responsible for the final grading of all the above projects.

(2) Industrial Specified Projects – Students can work on projects in cooperation with industry, but may not be paid for the work they perform for the duration of the project. This requires a signed agreement by the industry technical supervisor, the student(s), and the supervising faculty member. The industrial technical supervisor is expected to meet with the group weekly to provide technical leadership. Students on such projects will still be required to attend lecture and meet weekly with the supervising faculty to report on their progress. Here also, the supervising faculty will be responsible for the final grading. In certain instances, the supervising faculty may require that a university faculty member serve as a consultant in such projects.

(3) Student Specified Projects – Students can propose their own projects. However, a faculty member of the university must accept to be the nominal technical advisor for the group. To propose such projects, though possible, requires extensive lobbying by students.

Each design project must include the following components:

(1) A clear technical problem statement. Before proceeding beyond this point, each group should be certain that it has affirmative answers to the following questions:

- Is the problem solvable within the allotted time, say two semester or two quarters?
- Does the group have the expertise to complete the design, prototype, and testing?
- Does the group have access to the financing for the prototype?
- Does the group have access to the required test equipment?
- Is this a design problem (not research, nor fabrication)?
- Is the project significant enough to be worthy of the earned credits (requiring up to 12 hours work per week per person)?

(2) A quantitative performance specification.

(3) Quantitative analysis for possible design solutions, from which one solution is to be further developed.

(4) Detailed quantitative design of each component for the selected solution approach.

(5) Construction of a prototype.

(6) Development of a test plan.

(7) Evaluation of the prototype solution.

Each design must consider realistic constraints on prototyping and manufacturing costs, and per item consumer cost and pricing, as well as safety, reliability, aesthetics, ethics, and other possible social impacts of the design. Generally speaking, these are categorized into two groups, namely, **design constraints,** and *elements of design.* These two categories together is called *engineering constraints*.

Each project will have the following deliverables:

(1) Oral Design Review – Generally, this is at the end of the first semester or the first quarter. (We propose, however, to schedule this presentation for the beginning of the second semester or the second quarter, in order that students benefit from the extended break between two semesters.) The group presents the results of items 1-4 and a bill of materials with costs.

(2) Written final report – This is due during finals week. It should contain one section for each of the seven items listed above; a bill of materials; the realistic constraint analysis; and any schematic, block diagram, or other figure needed to describe the end product fully.

(3) Final design presentation and demonstration – The outline should be similar to that of the final report, but organized to cover the most important aspects within the time constrains.

We close this section by noting that clearly, in the forthcoming lecture notes, specific guidelines to present these deliverables are discussed, which those guidelines form the core of this course.

1.6. Senior Design Contract

A typical senior design contract is shown below, modifications are needed to use in different engineering programs.

Senior Design Contract

Senior Design I, II   Student Name:_____
☐ Individual   ☐ Team (members:_____
Technical Faculty Supervisor: _____ Supervisor Contact:_____
                                                                               (phone or e-mail)

Proposed Project Title: _____
_____
Laboratory, where the project will be performed: _____
Time and location of weekly consultation with technical supervisor:_____

Regulations

I. Meet with the technical faculty supervisor before the end of the first week of the semester to discuss the project.

II. Prepare a one-paragraph description of the project objective and expected activities, and attach that to this Contract form. Complete the Senior Design Contract, obtaining all necessary signatures, then submit the completed information to the course instructor. Note: Project teams need to submit only one contract for the whole group.

III. Prepare, in consultation with the technical faculty supervisor, a draft of the project specifications (about five pages) by the third week of the semester. Submit the draft to the course instructor for approval.

IV. Maintain weekly laboratory notes and weekly consultations with the technical faculty supervisor for the duration of the two semesters. Attend all weekly lectures delivered by the course instructor.

V. Prepare, in consultation with the technical faculty supervisor, a progress report and demonstration during the last week of the first semester. Submit laboratory notes and progress report for instructor review. A grade of in progress (or deferred) will be issued at the conclusion of the first semester indicating that the course is "In Progress" and will be issued a letter grade at a later time.

VI. Maintain weekly laboratory notes and weekly consultations with the technical faculty supervisor for the duration of the next semester (Senior Design II) as well. Attend all weekly lectures delivered by the course instructor.

VII. Make a formal presentation of the project in the last week of the second semester (Senior Design II).

VIII. Prepare, in consultation with the technical faculty supervisor, final project report and submit that to the course instructor by the end of the second semester (Senior Design II) as time and place will be announced.

IX. The final letter grade will be given to the second course (Senior Design II), and retroactively given to the first semester course (Senior Design I), replacing the in progress or deferred grade. Final grades will be based on timely submission of the abstract, the first semester progress report, the laboratory notes, the oral presentation as well as demonstration, and the final report. All endeavors will be made to achieve a **working project**.

Signatures

_____   _____
Student (Team Member 1)   Student (Team Member 2)

_____   _____
Student (Team Member 3)   Student (Team Member 4)

_____   _____
Technical Faculty Supervisor   Course Instructor

Copies will be provided to all team members and to the instructor. Also, a copy will be filed with the Department's Office of Student Affairs

1.7. Steps to Select a Project

Step – 0: Prepare a brief academic resume, which describes your specific technical strength and general background in less than two pages. It is very important that you make a case for yourself as why you should be doing a specific project. This step is more or less like applying for a job. Therefore this resume is the first draft of your future resume that opens a door for you.

Then follow one of the following steps 1A to 1C, depending on your situation.

Step – 1A: Make an appointment to meet and talk with professors with whom you wish to work, and see whether they are willing to recommend you for their projects. At least you should talk to two (preferably three) professors. This step is the same for all faculty members of the university. Or,

Step – 1B: If you have an industrial project in mind, which meets the above requirements as stated in the core of senior design, then you still need to talk to an engineering professor in your discipline. This professor must be willing to become your Technical Advisor, in addition to your Industrial Advisor. In many cases the professor in charge of this course may serve as a Technical Advisor, but not always. Or,

Step – 1C: If none of the above projects appeals to you, but you have your own ideas, then you must lobby for that idea with a faculty member, and see whether you can find a professor who is willing to become your Technical Advisor. This approach requires extensive efforts, but cab be done if it is planned in advance.

Step – 2: Identify one or two of your classmates who have similar interest and want to work with you on the same project and have gone through the same steps as you did. Then meet again with the faculty that all of you wish to work, to see if he/she is also willing to have all of you under his/her supervision.

Step – 3: Make a brief proposal to the course instructor regarding the title of project (and the professor's name), which you want to work on. In support of this proposal include a package consisting of your resume, your classmate(s) resume(s) if applicable, the professor's consent, if any. If you have already an endorsement of the faculty member include that as well. It is also required to have at least two more projects' titles in your proposal with supporting arguments, as your second and third choices.

Step – 4: Every effort will be made to match you with your best choices, although in certain instances changes maybe required. One consideration is an equitable distribution of senior design students among faculty members. In that case you will be notified promptly. Thus, have your contract ready for signature by your faculty advisor. You are almost done.

Part Two: Course Lecture Notes

Lecture Notes – A Summary

The engineering disciplines have their roots in mathematics and basic sciences, but through their own initiatives have evolved in many instances toward more promising areas of original scientific work than were possible in the traditional sense of mathematics and basic sciences. What makes an engineering education, however, different from its counterparts in the above areas is a sequence of engineering design courses, which are developed in these disciplines subject to a number of real-world engineering constraints. These design courses are intended to put students through a process of deliberating and considering a number of issues simultaneously as real engineers do in their daily lives. These design courses are based on a set of decision-making steps invoked repeatedly, in order to complete a design process, which entails establishment of a number of criteria that are called *essential elements* of these design courses. These elements, as motivated in part by the recent recommendation of the accreditation board for engineering and technology (ABET), are primarily as follows.

- Students are expected to establish technical objectives and goals.
- They are expected to establish methods of technical analysis and synthesis.
- They are expected to establish *a* method of technical construction and testing as well.
- Also, they are expected to establish scientific method(s) of technical evaluation and design comparison.
- All the above, must be accomplished within a specified time-frame, and subject to meaningful *engineering constraints* – as those faced by engineers in the real-world environment *concurrently.*

These design projects must encompass the following features:

- Development of student creativity.
- Use of open-ended real engineering problems.
- Development and use of modern engineering methodologies.
- Well-defined and well-specified design problems.
- Well-researched alternative methods of solutions.
- Well-evaluated final solution on a multiple-objective criterion.
- Well-constructed and cost effectively designed.
- Well-documented and presented final product in a real working or completed condition.

We have developed our lecture notes based on these general requirements and the fact that each engineering program must have a meaningful and major, thus non-fragmented, engineering design experience that builds upon technical, scientific, and social and behavioral sciences. Also, the fact that any engineering firm is inherently on the crossing road of many of these disciplines. We believe these notes are coherently assembled to cover aspects of these topics that are common to all senior design projects. We hope you find these notes useful, and that they will serve to provide a well-rounded engineering education.

The following list of Lecture Notes is developed to address these commonalties, as well as to emphasize that technical design is just a small piece of the "Big Picture". Namely, technical design as conducted by engineers is just *one* tile on the vast mosaic of an engineering enterprise. Perhaps the following table will convince us that an engineer must learn both social and technical skills in order to interact with a large group of professionals from an ever diverse host of people from all over the world. To interact with these people, and bring together the product to the market place in a manner to make sufficient income to sustain such an operation, is an art that we must learn, while we are still in school. To educate students about these concepts as they will face in their careers, is our goal. Looking at these matters and understanding the thought behind inquisitively is also a process that we hope to initiate in our students.

Table 1. The tiles on the mosaic of a typical engineering organization.

Human Resources	Technical Documentation	Marketing & Business Planning	Physical Plants / Facility Management	Risk Management
Engineering Research & Development	Intellectual Property, Patent & Legal	Investors And / Or Shareholders	Purchasing & Procurement	
Safety & Reliability,	Environmental Protection Agency / Other	Banking & Investment	Production And Manufacturing	Advertising
Quality Control	Government Liaison Offices		Accounting & Cash Flow Management	Shipping And Distributions
Ethics	Social Impacts	Public & Customer Relation		Sales

Lecture	*One:*	*Introduction and Course Outline*
Lecture	*Two:*	*Preliminary Issues, Expectations and Professionalism*
Lecture	*Three:*	*Final List of Projects and Contracts*
Lecture	*Four:*	*Laboratory Notebook and Proposals*
Lecture	*Five:*	*Proposals and Memos Continued*
Lecture	*Six:*	*Steps Towards Oral Presentations – Phase I*
Lecture	*Seven:*	*Technical Writing – Phase I*
Lecture	*Eight:*	*Marketing High - Tech Products*
Lecture	*Nine:*	*Marketing High - Tech Products (Continued)*
Lecture	*Ten:*	*In Preparation for the Design Review*
Lecture	*Eleven:*	*Ethics*
Lecture	*Twelve:*	*Design Review Preparation – The Final Thought*
Lecture	*Thirteen:*	*Mid - Course Design Review Results*
Lecture	*Fourteen:*	*Technical Writing – Phase II*
Lecture	*Fifteen:*	*In Preparation for the Final Report*
Lecture	*Sixteen:*	*Steps Towards Oral Presentation – Phase II*
Lecture	*Seventeen:*	*Engineering Economics*
Lecture	*Eighteen:*	*Entrepreneurial and Venture Capital*
Lecture	*Nineteen:*	*Entrepreneurial and Venture Capital (Continued)*
Lecture	*Twenty:*	*Quality Control*
Lecture	*Twenty One:*	*Intellectual Property Issues*
Lecture	*Twenty Two:*	*Intellectual Property and Patents*
Lecture	*Twenty Three:*	*Risk Management*
Lecture	*Twenty Four:*	*In Preparation for the Final Presentation*

LECTURE ONE: Introduction and Course Outline

Remark: In this lecture we review the *essential elements* of this course, followed by presenting a news item regarding Mr. Jack St. Clair Kilby – the 2000 Nobel prize co-winner in physics [1]. Next, an essay about the late Professor Thomas James Higgins [2], whose inspiring teaching is lingering in this textbook, is presented. Reviewing various sections in Part One is emphasized. It is expected that in a typical course setting the instructor will go over a few announcements and a possible list of faculty sponsored projects.

Special Resources:

[1] News Item, *Wisconsin Alumni Association,* Member Profile, Jack St. Clair Kilby, MS'50, p. 42, Fall 2000.
[2] In Memoriam – Thomas James Higgins: A Walking Encyclopedia, *IEEE Trans. Auto. Contr.,* vol. 44, no. 3, pp. 440-441, 1999.

Introduction

As stated in Section 1.5, under "The Core of Senior Design Courses," a list of projects as proposed by the faculty is available for students to select from, and a number of students have already done their research and know which project they want to work on. We believe that this trend should continue by allowing students to select their own project within a reasonable arrangement on the part of the course instructor. Here, we mean the students must, first of all, have a proper background to complete the project as explained in Section 1.7, under "Steps to Select a Project." Secondly, each team must be assembled coherently. The course instructor is certainly the final arbitrator in this selection process and should make the necessary adjustments to form teams and to match such teams with corresponding projects if needed. It is to be noted that each student should have planned this selection by contacting his/her faculty advisor, and in many instances by lobbying his/her ideas and choices well in advance, rather than as a manner of speech, in the last minute before graduation. It is also to be noted that if any faculty asks for particular students based on their earlier networking, then that faculty will get those students. The current senior students are expected to inform future students to plan ahead for this course. They should do so in their own special way of propagating words!

Course Outline

Students should be briefed on the scope of this course as outlined in Section 1.5, under "The Core of Senior Design Courses." Clearly, the course instructor may have his/her own structured syllabus for this class. However, due to the diversity of these projects, perhaps one rigorous syllabus for this course may not fit all cases. In many instances, this constraint is compounded by the fact that this course is constantly evolving, since, many of the commonalties are changing and often yet to be discovered. Therefore, an alternative and perhaps pragmatic way of handling this course is to set the standard, particularly regarding the course grading policy, in a manner that is fair and flexible, which upholds both the integrity of the institution as well as the basic rights of students. Thus, the course instructor is both a student as well as the institution advocate, and that is exactly why there is a need for a general course instructor to monitor all projects.

Grading Policy – An Overview

Although, each instructor makes his/her formal decision regarding the course grades based on the overall assessment of the projects, as measured, considering the following elements. The weighting of these elements that form various aspects of this course, may change from one project to another, thus requiring additional explanations beyond the allotted time during this first-hour lecture. That changing weight is the main difference between the administration of this course and other regular technical courses requiring special experience and wisdom on the part of the course instructor. Clearly, the course instructor reserves the right to make his/her final decision regarding the course grade based on the above fact.

(1) *The organization of the task,* which entails problem statement, approach selected to complete, the role and contribution of each team member.
(2) *The clarity of documentation to complete the task,* which entails explicit and detail logging of activity of each team member as described in this course, particularly Lecture Four, subsequently.
(3) *The quality and marketability of the task,* which entails description of novelty in both technical sense, as well as marketing of the final product, i.e., cost and related issues.
(4) *The superiority of the selected technical method,* which entails arguments on various technical matters including accuracy and repeatability of the design performance.
(5) *The intermediate assignments and reports,* which entails specific work to be assigned weekly pertaining to both specific project as well as some general issues discussed in lecture notes.
(6) *Final report and presentation,* which entails a formal report to be prepared in style and format that will be introduced in subsequent lectures. The final presentation must also be done according to procedures that will be described in class. In the final presentation issues pertaining to the underlying ethics and social impacts of the project will also be discussed.

Each of the above elements requires additional explanation and understanding. In essence these elements are topics that form the core of this course. Students should concentrate all their efforts in performing their duties as excellently as they can, and not to be concerned with weighting of each of these elements in their final grade. We will pay special attentions to those who show progress in learning and correcting their mistakes. We become more specific regarding each of the above in our future lecture notes. However, the final thought that we hope to convey to students is as follows. ***Excellence speaks for itself!***

Final Thought

The following news item concerns Mr. Jack St. Clair Kilby the 2000 Nobel Prize Co-winner in Physics, who is an electrical engineer, and in 1958, helped to invent the integrated circuit, or chip, which led to the foundation of modern electronics industry. The *essence* of this inclusion is to encourage students to set their standards and goals very high, and to realize that an electrical engineer can win a Nobel Prize in Physics. The picture in that handout shows Mr. Kilby had logged a drawing of his invention in his "Laboratory Notebook," perhaps not knowing that some day he will win the Nobel Prize. Also, in this news item Mr. Kilby is asked who is his favorite professor. He has replied: "T.J. Higgins. Back in 1950, UW (referring to the University of Wisconsin-Madison) instructor traveled to Milwaukee (where Mr. Kilby went to school) to teach graduate courses *several nights a week*, which is how I managed to earn an advanced degree while working full time." The essence of this reply is many-fold. *First of all,* most people who have succeeded and accomplished great things in their lives have done so by working very hard. Clearly, the Milwaukee of 1950's was not particularly very conducive to excellence in electronics research leading to a Nobel Prize. But Mr. Kilby worked very hard during the day to support his family and went to school at night to enhance his life to the point that we see the final result of his effort. Furthermore, this is one great model that we all must imitate.

Secondly, and perhaps equally importantly, Mr. Kilby mentioned Professor Higgins, who traveled several nights a week from Madison (about 80 miles west of Milwaukee) to teach at UW-Milwaukee, as his favorite professor. The author had the privilege of being one of Professor Higgins students as well. In this regard, an essay that is written about him, who passed away in 1998, is also included as *a point of reference.* It emphasizes how remarkably energetic was Professor Higgins, who taught at two different campuses, 80 miles apart, concurrently, and never complained about the work-load. The class must carefully read this essay and uncover all its subtleties and hidden between the lines remarks.

Fasten your seat belts, the expectations are very high and we demand excellence.

Closure

Generally speaking, the class ends with some additional remarks, some of which are documented in the subsequent lecture notes and are based on the mood of the day. Students are asked to finalize their choices by the following week and report their progress to the course instructor. We now offer a chart that we feel each student should review in conjunction with his/her project. These charts must be incorporated in the final report of this course.

Essential thoughts in this lecture

Issues.	Applicability to your project, if any.
Six elements for grading policy. (This is put here first to remind you that grades are given individually, not as per project or group.)	You may write, for instance, that certain parts of these elements do not apply to your project, because of certain reasons. Then present your rationale specifically on those elements. On the other hand, for those issues, which are relevant to your project and you have considered in your report, then explicitly highlight and describe those in your report. Please do not make any assumption that we know what you are thinking!
Team work, and project selection.	Please have your resume ready and form a group with right partners, then lobby for your choice as explained under "Steps to Select a Project."

JACK ST. CLAIR KILBY MS'50

(Reprinted with Mr. Kilby's permission.)

2000 Nobel Prize Co-winner in physics
In 1958, Kilby helped to invent the integrated circuit, or chip, that led to the foundation of modern electronics industry.

The big idea
Kilby first wrote about his idea in his notebook on June 12, 1958: (right)
"The following circuit elements could be made on a single slice (of silicon): resistors, capacitors, distributed capacitors, transistors."

A distinguished career
Kilby joined Texas Instruments in 1958, where he led semiconductor research and development and served as assistant vice president and director of engineering until his retirement in 1970. He then continued his career as an independent inventor.

Keeping good Company
UW-Madison's Dean of the College of Engineering Paul Peercy credits Kilby with the second of two watershed events in the miniaturization of technology. The first was the creation of the transistor, which was co-invented by fellow Nobel Prize-winner John Bardeen '28, MS'29.

Who was your favorite professor?
T.J. Higgins. Back in 1950, UW instructors traveled to Milwaukee to teach graduate courses several nights a week, which is how I managed to earn an advanced degree while working full time.

What advice would you give to grad students who are also working full time?
It's not that hard; it can be done

What are you most proud of?
My work on the integrated circuit and my family: two daughters and five granddaughters.

That being the case, have you worked to encourage more women in science?
Yes. I set up the Kilby Awards Foundation, which recognizes about six people a year for their excellence in science. Half of the award winners have been women.

What kind of music do you listen to?
Big Band music from the forties and fifties. There's a radio station here [in Dallas, Texas] that concentrates on it, so if you're careful, you don't have to listen to much else.

Have you enjoyed being named a co-winner of the Nobel Prize in Physics?
Yes, but I'm in some hope that the media attention will soon be over. I'll be traveling to Stockholm, Sweden, on December 10 to receive the award.

In Memoriam ---- Thomas James Higgins: A Walking Encyclopedia

(July 4, 1911 ---- September 11, 1998)

On September 11, 1998, Professor Thomas James Higgins, a giant in engineering education passed away in Madison, Wisconsin after a long bout with cancer. He was an intensely dedicated educator, and a person of exceptional humanity. His tireless devotion to learning and teaching in diverse fields was unparalleled. A voracious reader on an amazing range of subjects, throughout his illustrious life, he spent much of his time in the many University libraries as well as the Madison Public Library. He also devoured the content of numerous professional and popular periodicals to which he subscribed. He attended lectures and seminars in a wide array of disciplines, frequently participating knowledgeably in discussions far afield from his own main areas of expertise. He was truly a walking encyclopedia.

Although Professor Higgins' initial academic interests were in microwave theory and electrophysics, he was also very active in advanced automatic control theory during its early stage of development and remained so throughout his career. His interests in systems included advanced electric-circuit theory, large-scale systems engineering, and electric-machine theory, with major contributions in all the above areas. He was an early contributor to what has become known as the "boundary element method" in electrostatics and magnetostatics, his exact numerical computation of the capacitance of a cube solved a long-standing, classical problem. He was also an early contributor of applications of Schwartz distribution theory and helped to clarify and remedy shortcomings of various other known operator theory in circuits. He developed a general theory to determine the stability of linear *multiloop* servomechanisms that broadened the scope of, and yielded greater insight into, the control system operation [1]. During 1950's, he wrote a number of papers in various areas of control theory, which served as references for other prominent control scientists. One of his most important papers was "A resume of the development and literature of nonlinear control-system theory" [2]. During his career, which extended through his years as emeritus professor, he edited the manuscripts of over 120 textbooks in electrical engineering and associated areas for various publishing companies, published over 350 reviews and/or discussions of books and papers having to do with electrical engineering, mathematics and applied mechanics, was the author of more than 220 research papers published in various scientific and technical journals in the United States and Europe, and was the Editor or an Advisory Editor for seven journals published in the United States and Europe. He also wrote *Advanced Basic Automatic Control Theory* [3].

He was born on July 4, 1911 in Charlottesville, Virginia. He received, from Cornell University, the B.S. degree in electrical engineering in 1932 and the M.A. degree in mathematics in 1937. In 1941, he earned his Ph.D. degree (electrical engineering) from Purdue University. His teaching experience included: Instructor (mathematics) at Auburn Intercollegiate Center (1933-34); at Wyomissing Polytechnic Institute (1935-37); at Purdue University (1937-41); Assistant Professor at Tulane University (1941-42); Associate Professor at the Illinois Institute of Technology (1942-47); Professor in 1947-48. He joined the University of Wisconsin-Madison as Professor of electrical engineering in September 1948 and retired from active teaching, as Professor Emeritus, in June 1982.

Professor Higgins was a kind, considerate, and very thoughtful person, who unselfishly passed along information on topics which he knew were of interest to others. Once ensconced as a university professor, he sent thousands and thousands of letters and notes, some on scraps of papers, to scientists around the world, commenting on, or requesting copies of, their recent articles. In this process, he encouraged many young and emerging scientists, globally, in their early careers. During the peak of the cold war, he was engaged in a lively exchange of information with scientists in the former Soviet Union. This activity was a precursor and a catalyst to the establishment of the IFAC. Upon his retirement from "active teaching," he donated his technical books and the collections of his reprints to the engineering library. Shortly before he passed away, his voluminous correspondence and other papers were transferred to University Archives, and his collection of books of a historical nature to the UW-Madison libraries. Such acts were symbolically appropriate for someone who viewed the library as the *heart* of the university.

Professor Higgins received the *ASEE George Westinghouse Award,* in 1954; the *Benjamin Smith Reynolds Award* of the College of Engineering of the University of Wisconsin in 1963; the *Donald P. Eckman Senior Memorial Distinguished Activity in Education Award* of the Instrument Society of America, and the *citation of "Engineer of the Year"* of the Wisconsin Society of Professional Engineers in 1964; the *Certificate of Outstanding Service* of the American Automatic Control Council, and *Annual Appreciation Award* of the IEEE Systems, Man and Cybernetics Society in 1971; and an *IEEE Centennial Medal* in 1984. He won the *Edward P. Mikol Memorial Award* (for "Best Paper," at the 1986 ASEE Annual North Midwest District Meeting) and the *1996 Outstanding Professional Engineer in Education Award* of the Wisconsin Society of Professional Engineers. In 1996, he was also named an *Honored Member of the Wisconsin Retired Educators Association* in recognition of his outstanding service in education and dedication to the support of this association.

Professor Higgins belonged to 35 professional and honorary societies, held honorary rank in many of them (Fellow of IEEE, ISA, AAUP, AAAS, ASEE), and served many as officer or committee member. He was a registered Professional Engineer in Wisconsin and Illinois and was listed in most of the appropriate *Who's Who* volumes.

Professor Higgins' major professional avocation was the history of electrophysics and electrical engineering and the associated physical sciences with particular emphases on the life and work of the major workers in these domains. His often cited role model was Lord Kelvin, and he collected books and other publications about him. He published several comprehensive articles in these areas and believed in exposing students to the historical background of scientific developments. He stated his rationale for zealously pursuing this matter in "The history of the development of electrical engineering and electrophysics," with over 1100 references [4]. There he quoted these words from Macbeth: "Look into the seeds of time, And say which grain will grow and which will not."

Professor Higgins supervised 55 Ph.D. theses and over 147 master theses in electrical engineering during his active teaching years. Also, he adjudicated at least 94 Ph.D. and D.Sc. theses from various universities in India and Australia (the last one early in 1998) and was a member of the Selection Board for Appointment of Professors and Associate Professors, University of Bangladesh at Dacca, during 1971-75. Perhaps his main contribution was that of having led and supervised this extremely successful group of Ph.D. students, who in their turn supervised several hundreds of Ph.D. theses, and counting, in engineering, mathematics, and computer sciences----all within one man's lifetime----and this propagation is an extraordinary achievement.

One of the most remarkable characteristics of Professor Higgins was that he had never acted as a rubber stamp for any matter before him. He did his thinking independently. He was always the most reliable reader of any document, irrespective of its size and/or complexity. His powerful faculty was not accidental, but rather the result of a lifetime striving to be focused, and a desire to penetrate to the heart of an issue for a critical evaluation. He never took any matter for granted, but studied it carefully, and always respected those who were doing original work by recognizing their efforts. He was known to be very crisp. For instance, during the discussion about the Transactions being too theoretical, he entered and gave the following quotation from Boltzmann: "Es gibt nichts Praktischeres als eine gute Theorie (There is nothing more practical than a good theory)," which put the critics on notice!

Academically speaking, he was my grandfather and that was a very special relationship, resulting in many memorable encounters, in which I immensely benefited from his generous advice. Once in 1974, confronting a personal difficulty, he told me "be optimistic and don't let anything bother you." He was the ultimate optimist who enjoyed life to the fullest and had the good fortune of shaping many lives. He was unforgettable!

Professor Higgins is survived by his second wife Professor Emeritus Mary Ellen Roach Higgins, his daughter Professor Janet E. Higgins, his son Mr. James L. Higgins, an engineer with Boeing, and his brother Mr. Francis R. Higgins, retired.

Many people have wholeheartedly contributed to the preparation of this tribute, in particular his former Ph.D. students: Professors G.W. Swenson, T.A. Estrin, J.J. Skiles, A.T. Tiedemann, R.S. Marleau, and Vidyasagar as well as some of his friends Professors B.R. Barmish, W.R. Perkins, E.I. Jury, M. Jamshidi, and J. Zaborszky. Thank you all!

With very fond memories

Mansour Eslami
The University of Illinois at Chicago
Dept. of EECS (M/C 154)
Chicago, IL 60607-7053

References

[1] AIEE, 1954.
[2] *Trans. ASME,* 1957.
[3] T.J. Higgins, *Advanced Basic Automatic Control Theory.* Madison, WI: Madison Press, 1954, 346 pp.
[4] The Bulletin of Bibliography, 1952.

LECTURE TWO: *Preliminary Issues, Expectations and Professionalism*

Remark: The purpose of this lecture note is to touch upon very briefly issues and lessons in professionalism. Also, to brief students on the status for selections of their technical faculty advisors and technical projects. Students have been advised and indeed highly recommended to review Section 1.6, under "Senior Design Contract," in order to understand the meaning of entering into a contractual agreement with other parties.

Introduction

To finalize students' selections is a major task for everyone associated with this course. Thus, students are to be briefed on the status of their selection, which some were submitted a few days before. Those who have not done so are reminded to speed up their selection, and allow us to begin the work. Copies of "Senior Design Contract," should be distributed in class in order that students fill out this form and be ready to receive their faculty advisors' signature as soon as the list of projects is final. In other words, as soon as we know who is working in which project, these forms should be ready for the course instructor's signature. Without that signature, we really cannot begin to work. These announcements are posted on the instructor's office door, or on the course web site. To avoid any possible hard feeling, we normally do all these selections as openly and participatory among students as we can, and that approach takes more than half of the class time. We have learned quickly that under the best conditions it takes about three weeks to finalize the list of projects and nothing can be done formally until that list is constructed. In order to proceed, however, we may assume that all projects are assigned, and all administrative matters are completed duly.

Preliminary Issues

Since almost all projects are performed in a group, the first thing that comes to mind and we must be concerned about is how do we work in a group? This situation is different from your earlier laboratory courses, where you had worked in a group of two or three. Here, the stake is much higher than before. The selection you make may affect your life for many years to come. For instance, during your job interview, you will be asked about your senior design project, because that is the most likeliest topic you will be asked in order to break the ice. You may even get your first job based on you senior design experience. If you continue your graduate study you may select a topic similar to what you have done in your senior design project. Thus, the structure and discipline that you get in this course will be with you for a long time. Based on all these factors and many others that constitute what is known as experience, our response to the preceding question, which you should take that with yourself for the rest of your life, is as follows.

"Protect yourself."

This warning may surprise and possibly confuse some of you, and you may ask what happened to all that team-work discussion? One implication of the above is as follows. Why do I need to worry about my friends? The fundamental answer to all of these questions is as follows. You must protect yourself in order to protect your team members or your friends. Because,

(1) No one needs a careless and unengaged friend or team member. In other words, your preparatory

and final work must be so complete that if your team member loses his/her notebook or records, then you can fill in right away.
(2) Take nothing for granted, and make no assumption, and rely on no one but yourself.
(3) Be intimately familiar with all facets of your project, and do not let anyone do the thinking for you. When you allow others to do your thinking, you lose immeasurably and immensely.

If you follow these not so easy steps, then you are guaranteed success in all your projects. But if you choose to remain passive and not to take charge of your affairs, then you are constantly behind. We give explicit guidelines on how to develop and manage your data and logging of your activities subsequently.

Expectations

In just a few months, you will become an engineer and be expected to act in certain ways, which are very different from being an undergraduate student. As an entry-level engineer you must be concerned about your boss as well as your client, although, in many instances, you may not directly face a client right away. It is expected from you to be precise, which in turn means you should learn to avoid side issues when you talk to your boss. Also, be respectful of your boss's time, which requires when you meet your boss to have a pen and a note pad ready and record your conversation. This courtesy you extend will be appreciated, and furthermore by taking note and not relying on your memory, you also force your boss to be accurate and do not give you a runaround. Do not walk into your boss's office with your backpack and waste his/her time by trying to find a pen or a piece of paper to record your conversation or to ask a question. The more professional you conduct yourself, the more results you get. In other words, do not give any excuse to anyone to deny you guidance and assistance.

Now, your faculty advisor is your boss, and we do not want to hear from anyone of them that you went to his/her office unprepared. Act like a professional engineer.

Make a complete list of your activities as far as your project is concerned. We give specific guidelines as how to assimilate your bookkeeping in future lectures. For the time being get ready to write a list of your interactions with your team members and keep a minutes (record) of your conversations.

Soon you will be asked to deliver specific assignments to both your faculty advisor and the course instructor. These deliverables must be prepared with certain format and styles, which bears yours and your team members' names, date of submission, almost similar to your laboratory report until further notice.

Professionalism

To put the above in a slightly different setting we suggest that in your daily conduct as soon to be an engineers try to portray yourselves as *professionals* – that elusive word which has no globally acceptable definition, although the following list may qualify for one such definition. A professional person – in this case engineer – conducts his/her business in a manner that:

- Commands the respect of his/her clients, because they trust his/her judgement.
- Commands the respect of the society for his/her profession.
- Balances the quality of his/her work versus the need to succeed his/her business – An issue of fairness and reciprocity, which is the cornerstone of commanding respect.

Extra effort must be made to convey the above image of professionalism. In many occasions you must sadly accept the fact that others do not have the same level of commitment and you must adapt to that situation quickly. One cannot be professional in any profession, if he/she does not value education and training, which are prerequisites to command respect of others. Also, if a person does not adhere to the ethics of the profession. Nor, if a person does not make strong efforts to portray a healthy image of his/her profession. There are many studies, which show certain professions are not respected by the society at large. Engineering is not in that list and we are committed to keep it that way.

Professionalism puts all the above together and keeps us moving forward. There are more specific guidelines on how to be professional, which are covered throughout this textbook. The simplest definition however, is our decision-making practices in technical and business procedures that identify us as professional or otherwise.

Final Thought

Be particularly concerned with the notion of a client for whom you are performing a task as an engineer, and depend for your livelihood. (In an ironic sense your faculty advisor is also your client who may recommend a good grade for you). Almost always times play a paramount factor in maintaining the client's interest in the ultimate product of a project. Never ever procrastinate even when you have a signed and sealed contract and you are ahead of schedule. Contract means nothing when the client loses his/her interest. You just do not want to spend your time in any court of law to litigate a contract, even when you are one thousand percent right. Cost and timeliness of a performed task are as important, if not more so, as technical matters associated with the project. Also, do not overestimate your client's knowledge in your duties and procedures. However, be aware that claims have been made by clients, alleging that they have not been properly explained certain procedures when a product has been delivered to them. Be both careful and cautious in dealing with your clients. ***Do not lose sight of these matters.***

Closure

The class ends by reminding students to get signatures from their faculty advisors as soon as possible, in order to proceed the tasks ahead.

Essential thoughts in this lecture

Issues.	Applicability to your project, if any.
The importance of the bookkeeping of main activities is discussed. Formats and procedures yet to be explained. Professionalism in several forms is discussed.	Obvious!
Do you want to add anything else?	Please elaborate.

LECTURE THREE: Final List of Projects and Contracts

Remark: In this lecture issues pertaining to Senior Design Contract and the final list of projects are described. Perhaps this is the best justification for the use of the word experience in the title of the book since this is one important place, which our experience in setting up this course is reflected.

Introduction

It is expected that by this lecture all issues pertaining to the final selection of the projects have been resolved and the list of projects and individual copies of signed senior design contracts are ready for distribution among corresponding students. Please recall that we have already reviewed the contract with students. The quick resolution of the selection process is only possible as the results of the efforts by both students and faculty. Before closing this matter, we propose to give one final chance to everyone who may possibly have a second thought and may wish to change his/her project. If there were no objections and everyone concurs to proceed unanimously, then we do so. Otherwise, we need to take care of the selection matter urgently. Students should be reminded that the signed contract must be kept as an important part of their record. Students should also be advised that this sort of paperwork is an inevitable part of the engineering practice and we all need to get used to that routine.

The class should be briefed on the fact that the work must start immediately, if it has not already been commenced. They should be reminded that every effort must be made to complete the task well ahead of its deadline completely and duly.

Looking back at this selection experience, one wonders whether we could have saved at least one week. Perhaps we could, if we had distributed the list of projects sooner, but that had other implications, which we do not want to get into, and each instructor has to decide how to plan in order to maximize the outcome of this endeavor. Certainly, the above should be considered as a list of suggestions. Our experience is that we always run into a few students who could not get all the needed work done in a timely manner and they will slow the rest of the class inevitably. We have always reminded our students that they should not expect a long break, and instead they have to work continuously until the project is done.

Reflection

Let us pause for a moment and ask what have you committed yourself to? You have committed yourself to carry on a design project meriting four to eight credits on your own. Are you equipped to do so? For those few who have planned the above selection in advance, they perhaps have a well-defined problem to work on and we give some general guidelines to continue. But for many of you who have just joined a group or select a project, you need time to evaluate the project and with the help of your technical faculty advisor to begin the process of research and the preliminary analysis to sort out various issues concerning your design project. We are fully aware of the diversity of various engineering projects and the fact that one approach to tackle the problems does not fit all. Some of our projects need additional research and learning before they become clear to work on. That is why we could not endorse one rigid syllabus to complete all projects. However, we try very hard to guide all projects in a manner that within the allotted time we get the best possible results.

Having said the above, we propose the subsequent organization in order to continue the work. But first, we call a **well-defined project** the one that has a well-defined problem statement and also has a corresponding motivated group of students who have already commenced the work. Similarly a project that requires additional research and conditioning, which conditioning refers to both its problem statement and its corresponding group of students who may need preparation, is called an **ill-defined project**. We can *model the difference* between these two projects *by time* if we are very optimistic, but time is not the solution for all cases. However, time is also something that we have very little to spare, nevertheless we are obliged to offer some and we do so. But please put your act together as soon as possible.

Task Organization

Whether you have thought about your project or have begun to get into it, you must organize your work in the following general manner. Specific details will follow in our future lectures. For the time being keep a copy of details of all your activities and interactions with your team members until we give specific instruction to document those. Also, keep the following thoughts in mind as venturing in uncharted areas of design and creativity.

(1) Draw a flow chart of steps you need to take to finish the task. For instance, write in your own words the problem statement as you understood it, and how you want to work on that. Check with your team members to see how they react. Try to give a qualitative and quantitative solution of the problem, and see how you can divide the work among yourselves in order to finish the job.

(2) Roughly, try to figure out what will you need, as far as laboratory space and equipment is concerned? How much materials, supplies, *etc.*, are needed to complete the job.

(3) Try to estimate the overall cost of your project, although you have not really chosen all details and do not know which specific solution, if more than one, you will be using. Certainly, this is hard, but we put it here to remind you that you need to work with your technical faculty advisor to choose a solution that you can afford. You will be hearing very often in industry the word *affordability,* which means the sky in *not* the limit! You must have a very good idea about the cost of production and testing and how to secure funding for those activities. Keep track of all your purchases as you move forward. You need to keep the receipts for all your expenses throughout this course. If your department is going to reimburse some of these expenses, no commitment is proposed, then they will ask for receipts.

(4) Finally, put all these rough ideas and estimates together and share that information and thought with your team members and your technical faculty advisor in some sorts of memos to them. Soon we give specific format and structure to prepare these memos formally, with a copy to the course instructor. In most cases your client will dictate his/her specific style and approach to prepare these memos. But remember what sells *mostly* is the substance and not style. Thus, for now you do your substantive work, and later we show you how to package it with style!

Final Thought

In order to make this an enjoyable experience, please accept some of the changes, which we suggest that you need to incorporate in your routines. You need to work at least 8 - 12 hours per week on average (depending on the course credits), if you are expecting a good grade. Please work with your course instructor to plan this activity right!

Closure

If the above is in order and students are on the right track to carry out their tasks, then the class will be ended with a sense of relief. However, if that is not the case, then the instructor should make every effort to settle these matters as quickly as possible. The class should also be briefed that soon they would be asked to give an organizational flow chart or approach for their projects.

Essential thoughts in this lecture

Issues.	Applicability to your project, if any.
Organizational matters.	Obvious!
Do you want to add anything else?	Please elaborate.

LECTURE FOUR: *Laboratory Notebook and Proposals*

Remark: In this lecture we emphasize the procedures that students must follow to record their daily activities pertaining to senior design projects. These activities should be regularly archived for a possible future need to support an intellectual property or patent application. Students are briefed on how to prepare proposals and memos in order to report their progress.

References:

[1] M. Eslami, *Analog and Digital Circuits, Theory and Experimentation.* Melbourne, FL: R.E. Krieger, 1987 (pp. 22-23).
[2] H.M. Kanare, *Writing the Laboratory Notebook.* Washington, DC: American Chemical Society, 1985.

Introduction

As much as we like to say all set and ready to carry out the technical assignment, we realize that there are many other procedural matters that must be learned before proceeding further. Students are briefed on the procedures, which they must follow to collect data and to structure their bookkeeping and logging of their activities and thoughts, in order to conform to the corresponding legal standard. This information collecting is followed by some suggestions on how to begin preparing proposals and memos to report their progress.

Laboratory Notebook

Generally speaking, the so-called laboratory notebook is a complete log or record of all work done by the experimenter. While there is no universal set of rules for keeping this notebook the following guidelines should be helpful. These suggestions on collecting data in laboratory are gathered from a textbook by this author [1], and more can be found in [2]. Subsequently, we refine these suggestions to include the logging of activities. In most cases a Laboratory Research Notebook, which is available at the Campus Bookstore may be used for this purpose. Although in many instances and as far as this course is concerned, this notebook can be replaced with a cheaper one that is marked by the student in the format described below. At the end of each session with your team member(s) you are required to pass a raw data sheet of your work to the other person(s) for his/her record. Thus, the issue of a medium to record activities and generated data is not cosmetic, but rather substantive, and by Laboratory Notebook we really mean a warehouse of information. You can build this warehouse on any medium where information can be *stored and retrieved,* in conjunction with a physical warehouse for your tangible records.

Having stated the above, we acknowledge that when we set our sight high and anticipate greatness – maybe a patent or an intellectual property down the road, we must invest time and follow a common-sense approach to archive our research and discovery as legal professionals advise us to do. For instance, when a patent examiner is reviewing an application, the actual ***date*** of one's discovery is the single most important factor in the final outcome. To establish ***when*** we had accomplished a discovery, we need to rely on our notebook, which often is challenged in the court of law by our competitors. Any irregularities in our bookkeeping system become red flags to our opposing parties. Many of us feel burdened by strict bookkeeping practices and feel that is unnecessary. However, a solid system of bookkeeping saves us valuable time, not only to protect us in a legal setting against allegations of conflict of interest or research fraud, but also in many other instances of accidental fire and similar loss of data or other valuable information. In the following, and for the sake of discussion, we assume that our Laboratory Notebook is a bound professional type that is designed for this purpose. Perhaps we should clarify that by a *laboratory session,* we mean any gathering of a research group or technical team for the purpose of deliberating or any kind of work on the project. Also, any gathering of individual participants is called a laboratory session – sometimes this gathering may take place at the library or a conference room.

A. On Collecting Data

(1) *Entries in the notebook should be in chronological order.* Pages should *not* be skipped in the body of the notebook in order to be used at some later date. Each page of the notebook should be numbered. The experimenter's name, date, section, and the names of any team member(s) or collaborator(s) should appear on each page. The notebooks to be used, or constructed (check with your course instructor), in this course must have the pages pre-numbered. There should be no blank spaces (except margins) on a finished page, otherwise cross the unused portion of the page with an "X".

(2) All information should be recorded directly into the notebook as soon as it is obtained. Memory or scraps of paper should not be used for a primary recording medium. The risk of loss or error is too great. The experimenter should be in possession of his/her notebook whenever he/she is working in the laboratory. Rough calculation and initial conclusions, should be recorded directly in the notebook.

(3) Do not crowd entries. Notebook paper is undoubtedly the least expensive item in any experimental investigation. Leave ample margins at the top, bottom and sides of each page for reference notes, which may be added subsequently, and if so those additions must be initialed, dated, and witnessed. *Entries should be in ink or ball point pen.* If possible use the same pen throughout the day, in order to support the idea that entries were all made at the same time and not altered later. *Never erase or obliterate.* Cross through an entry if you must, but leave it legible. Only immediate errors of entry should be crossed through. Observations that are subsequently found to be in error should never be crossed out; the fact that they are in error will be recorded on a subsequent page and a margin reference to this page can be inserted. Finally, be sure that your writing and numerals are legible and understandable.

(4) There are some records that cannot conveniently be put into a Laboratory Notebook, such as long strips of recorder paper or a stack of Polaroid prints. Such records should be immediately marked with identification numbers and logged, together with information on their source, into the Laboratory Notebook. Identification systems need not be elaborate, but they must permit unequivocal access in both directions between the Laboratory Notebook and such separate records. The identification system must satisfy any labeling requirements for specimens or supplies. A beaker, bottle or box used in the experiment should either be empty or contain a label identifying its contents and recorded in the Notebook. Keep your Notebook in a safe and fireproof cabinet, and ask all those related to the project do the same.

B. On Logging Activities

These entries are also should be made in the same Notebook as above.

(1) Each entry should document all you have done in each laboratory session. You must also include all pre-laboratory work or any sort of preparation that was put into the meeting individually and as a group. All entries must be made in the form of a diary and not a memoir, by reporting things as they happen and not as they recall.

(2) All meetings should be reflected in every member's Notebook similarly. One meeting per week is a must. All decisions agreed upon must be reflected in each member's Notebook, with explicit defined and agreed assignments for each individual member. There must be a well-defined timetable as how these tasks are to be completed. If failed to meet the deadline, then describe why, and how that task is rescheduled. In other words, we want a clean operation.

(3) While conducting an experiment, whether testing a circuit or anything else, draw a clear schematic diagram of the work to be done, and provide an analysis of the experiment as appropriate. Leave room for some sorts of comparison as you advance in the course.

(4) A major part of your activities concern locating and discussing issues with people outside the company (here university). Always write the name and telephone number of the person you spoke with and the date of this conversation. Write the key issues that you spoke with the person. For instance, if you have contacted a vendor, write the name and description of the component or item that you have discussed and want to purchase. Get a confirmation by asking the outside person to send you an e-mail about the discussion. In many cases, they tell you that they are busy, so it is also a good idea to get the person's e-mail address and the Fax number, in order that you send him/her a confirmation note. You will be surprised how often people change their mind without telling you!

C. On Logging Thoughts

As stated before, a useful Notebook must look like a diary that reports what you are thinking and doing, and *not a memoir*. It must display events as they occurred, and it must be easy to understand and reproduce. It must convey the evolving thoughts that are the cornerstone of research and future discovery. It must explain your planning, preparation, and expectation of the results. In other words, if it cannot be shown that there was a grand scheme to reach a goal, then the experimenter would have difficulty to prove his/her case in the court of law when seeking a patent or any other recognition. The description must be so lucid that years later someone with similar background can pick up the Notebook and understand what you were thinking and doing then.

The Laboratory Notebook or Journal as is called in the contemporary era must be a living and dynamic piece of work reflecting the thinking and effort which have been put into the project, and a complete awareness of the potential discovery from the project.

The Laboratory Notebook is also a barometer of how much time and effort has been put in a project, and therefore can serve as a reminder to us whether we are utilizing our resources wisely or not. Be aware that a documented life may also come to haunt us!

D. On Electronics Media

Many students would prefer to keep their records on their computers and produce daily or weekly journals. All we can add is that the burden of proof is on their shoulders. They must substantiate the originality and date of occurrence just as anyone else. One way to archive these notes is to have them on a disc or tape, with date, and *mail* that to an impartial party for safe keeping. Normally, the computers carry their own time stamping programs and automatically date the creation and editing of files, but still you need to prove.

E. On Archiving Laboratory Notebook

To verify the legitimacy of scientific claims made based on information recorded in a Laboratory Notebook, one must seek on a regular basis the witnessing and reviewing of an impartial or disinterested party. For instance, two colleagues who do not necessarily work on similar projects may review each other's work in confidence. The frequency of this verification depends on the nature of the work. For a fast breaking research this verification may occur on a weekly or even a daily basis, but the key issue is that the person who verifies must be trustworthy, and therefore must be reputable in the community at large. The nature of this verification may be both actual physical observation of the work done, as well as technical review of the outcome of the work. In case, the witness or reviewer suggests any correction or procedural improvement, that must be incorporated in future entries of the Notebook with a proper declaration of the reviewer's input. There is a fine line as how much a reviewer can play this role or becomes an active member of the team. If the reviewer crosses that line then becomes a participant in the project and must be replaced by another impartial party. Thus, select a reviewer who can do the job for the anticipated duration of the project and without any future problems. Team members cannot serve as each other's reviewer. An alternative to having a reviewer is to photocopy and videotape the research results with actual dating and *mail* that to an attorney for safe keeping. The mailing date of the package will stand when opened in court.

Whatever you write in your record book, in any shape and form, must display your thoughts and dreams and how you got the ideas and what you are looking for. It must be easy to read and easy to understand. Finally, it must be verifiable.

Preliminary Issues to Prepare Proposals and Memos

We will give details on this matter in our future lecture notes. Meanwhile, we want you to realize that a major part of your activity concerns literature search and review of published work in the area that you are concentrating. Equally important is browsing the web. When you are conducting these searches write down the key information about each such entry in your Laboratory Notebook. In this regard, include a brief description of the key issues as you see it. As you gain experience, you may not agree with these initial assessments, but that will be decided later. Now, as far as the citation is concerned, when you want to refer to an article that you had read in one of the IEEE Transactions, *per se,* you must follow the standard procedures described in the two attachments (one for the IEEE style, and the other for the ASME). Thus, write all the pertinent information for articles, books, URL's of web sites visited in the format selected by your course instructor before you forget. We pay special attentions to these efforts.

The following format for your *intermediate proposals and memos* should be observed.

(1) The total length should not exceed four double-spaced, typewritten, stapled pages.
(2) The allowable font size is 12 points.
(3) Upper and lower margins (includes gutters) should be 1.5 inches. Left and right margins should be 1.0 inches.
(4) The first page should start with an appropriate title. Each page should have your name in the upper right-hand corner. This may be included as a header.
(5) If this is a group memo, then everyone's name must be included.

Final Thought

We now have the essential tools to organize and document our work in a solid structure of academic and industrial settings, and in due time we elaborate on improving these matters as we proceed in our endeavors. The road ahead is certainly bumpy, but we have been there and done that!

Closure

The class is reminded of the pending progress reports from students due next week.

Essential thoughts in this lecture

Issues.	Applicability to your project, if any.
The importance of preparing a Laboratory Notebook and its role in this project oriented course with all details.	Just look at the illustrious picture of Mr. Kilby in Lecture One. The old saying that a picture worth a 1000 words could not be any more relevant than here.
Citation styles and formats for proposals and memos.	Obvious!

Appendix A:

Attached is a set of sample forms for citation styles, courtesy of friendly staff at Science Library of the University of California at Riverside a wonderful place to study.

IEEE Citation Style

In Text Citation
Citing in text is done with the [#] format. Punctuation occurs after the notation.

Example: A preliminary model was analyzed in a series of off-line computer simulations [1].

List of References
References are listed in the order that the articles appear in the text.

Journal Articles:
[#] Author's initials authors last name, other authors, "Article title," *Journal Name (typically abbreviated)*, vol. #, no. #, pp. #-#, Date.

Example: [1] M. Rucci, G. Tononi, "Registration of neural maps through value-dependent learning," *J. Neurosci.*, vol. 17, no. 1, pp. 334-352, Jan. 1997.

Books:
[#] Author's initials last name, other authors, *Title of Book*. Place of publication: Publisher, date.

Example:
[2] D.O. Hebb, *The Organization of Behavior*. New York: Wiley, 1949.

Article in a collection:
A.J. Albrecht, "Measuring Application-Development Productivity," *Programmer Productivity Issues for the Eighties*, C. Jones, ed., IEEE Computer Soc. Press, Los Alamitos, Calif., 1981, pp. 34-43.

Article in a conference proceedings:
M. Weiser, "Program Slicing," *Proc. Int'l Conf. Software Eng.*, IEEE Computer Soc. Press, Los Alamitos, Calif., 1981, pp. 439-449.

Do not include the editor's name for a proceedings unless it is carefully edited and published as a regular book

Dissertation or thesis
B. Fagin, *A Parallel Execution Model for Prolog*, doctoral dissertation, Univ. of California, Berkeley, Dept. Computer Sciences, 1987.

Electronic publication
L.P. Burka, "A Hypertext History of Multiuser Dimensions," *MUD History*, http://www.ccs.neu.edu/home/home/lpb/mud-history.html (current Dec. 5, 1995).

Patent
M. Hoff, S. Mazor, and F. Faggin, *Memory System for Multi-Chip Digital Computer*, US patent 3,821,715, to Intel Corp., Patent and Trademark Office, Washington, D.C., 1974.

Technical report
C. Hoffman and J. Hopcroft, *Quadratic Blending Surfaces*, Tech. Report TR-85-674, Computer Science Dept., Cornell Univ., Ithaca, N.Y., 1985.

Technical or user manual
Unix System V Interface Definition, Issue 2, Vol. 2, AT&T, Murray Hill, N.J., 1986.

UCR Science Library

ASME Citation Style

In Text Citation - Within the text, references should be cited by giving the last name of the author(s) and the year of publication of the reference. The year should always be enclosed in parentheses; whether or not the name of the author(s) should be enclosed within the parentheses depends on the context. The two possibilities are illustrated below:
It was shown by Prusa (1983) that the width of the plume decreases under these conditions. or
It has been shown that the width of the plume decreases under these conditions (Prusa, 1983).

In the case of two authors, the last names of both authors should be included in the citation, as shown in the above examples, with the word "and" separating the two authors.

In the case of three or more authors, only the last name of the first author of the reference should be included, as shown in the above examples, with the other authors being denoted by "et al."

In the case of two or more references with the same author(s) and with the same year of publication, the references should be distinguished in the text by appending a lowercase letter "a" to the year of publication of the first cited, a letter "b" to the second cited, etc. The references should follow the examples shown above.

List of References - References to original sources for cited material should be listed together at the end of the paper; footnotes should not be used for this purpose. References should be arranged in alphabetical order according to the last name of the author, or the last name of the first-named author for papers with more than one author. Each reference should include the last name of each author followed by his initials.

(1) Reference to journal articles, papers in conference proceedings, or any other collection of works by numerous authors should include:
year of publication
full title of the cited article
full name of the publication in which it appeared
volume number (if any)
inclusive page numbers of the cited article

(2) Reference to textbooks, monographs, theses, and technical reports should include:
year of publication
full title of the publication
publisher
city of publication
inclusive page numbers of the work being cited

In all cases, titles of books, periodicals, and conference proceedings should be underlined or in italics. A sample list of references in which these forms are illustrated follows.

Sample References

Sparrow, E. M., 1980a, "Fluid-to-Fluid Conjugate Heat Transfer for a Vertical Pipe - Internal Forced Convection and External Natural Convection," *ASME Journal of Heat Transfer*, Vol. 102, pp. 402-407.

Bejan, A., *Heat transfer*. New York : John Wiley & Sons, Inc., c1993.

Tung, C. Y., 1982, "Evaporative Heat Transfer in the Contact Line of a Mixture," Ph.D. Thesis, Rensselaer Polytechnic Institute, Troy, NY.

Most of the information on this page was taken from:
(No date). References [Online]. ASME. Available: http://www.asme.org/pubs/authors/ [1999, April 9].

LECTURE FIVE: Proposals and Memos Continued

Remark: The purpose of this lecture note is to clarify issues that were explained before and that students are still having problems to implement. In this lecture we begin to ask students to describe their experience so far and share some of their concerns in class. For many of them, this is the first time they talk in front of a group.

Introduction

Students are briefed on some of the earlier issues, which were discussed in class such as the importance of the Laboratory Notebook and related topics. Students should be invited and indeed encouraged to talk about their projects and some of their experience, as a sign of what the rest of the class should expect soon for their mid-course oral presentations. At this time, there must be quite a few questions among students, which show that they are beginning to engage themselves in this course. Otherwise, there is something wrong and we must find what that is.

Proposals and Memos Continued

Based on some feedback from the class we should continue to emphasize the importance of a proper presentation of any document or project. Clearly, students are interacting with their technical faculty advisor and definitely have written a few memos or proposals about their projects. Although, we have not yet given any structure and style, they are ready to see the first structure for a technical document. In other words, they all have a story to tell, but do not know how. Here is our first attempt to structure these documents.

In this regard we expect that the students pay particular attention to the following issues. Each document must be easy to read, giving a brief introduction *without any acronym*. Then start explaining your problem by walking through **each step** and gradually presenting your overall goals. Do not give the "Big Picture" first, nor assume that people who read your document (proposal) are intimately familiar with your work. In fact they are not! Also, we elaborate on this "Big Picture" subsequently.

Now, going back to the preceding sentence, if you really have read it thoroughly, and understood the true meaning of **"each step,"** then we have nothing else to offer you! In other words, the course would have been over and we all could have called it off. However, with all due sensitivity and respect, we believe that, it will take decades for someone to appreciate the meaning of **"each step,"** and be able to sort out his/her thought in order to write that way. In other words, to compartmentalize a thought in several steps is a multi-year task. But that does not mean we should be discouraged, and do not even try, not at all. Sooner we start, sooner we learn how to write correctly and effectively, which not only describes an issue clearly. But also, it gives other subtle information, which is hidden to a novice, nevertheless is imbedded in the text of the document. We will be looking into some such writing as an exercise in class.

Finally, we all have heard that we must write such that we are understood. That is not enough in this class! We must write such that we are ***not*** misunderstood, and that is not too much to ask.

As a first step toward organizing your proposals, we suggest that in your writing you pay special attention to the following items.

A. Technical

(1) Block Diagram, can you use that to simplify your presentation?
(2) Parts List, put all that with the corresponding specifications into one group.
(3) Special Software, sort these out accordingly.
(4) Special Resources, this is also required to be looked at by each project separately.
(5) Test Plan, clearly you need to look at this issue very seriously.

B. Administrative

(1) Budget, you must address this issue.
(2) Personnel Assignments, each team member has specific job and that should be detailed.
(3) Schedules, you should give a time frame to complete the task in hand as you see it.
(4) Non-compliance, you must discuss about each and all that went wrong among you!

C. Lessons Learned

Things will go wrong more often than you think. Simply state your experience when any or all the above go wrong and write whatever you have learned from this experience. A project report without lessons learned is not believable and is considered weak.

Final Thought

This is the first structure for submitting your documents. We hope to help you sort your presentation out and work with you to improve your writing skill.

Closure

The class ends by emphasizing the importance of compartmentalization in any presentation.

Essential thoughts in this lecture

Issues.	Applicability to your project, if any.
Preliminary elements that should be included in each proposal.	Obvious!
Do you want to add anything else?	Please elaborate.

LECTURE SIX: Steps Towards Oral Presentations – Phase I

Remark: In this lecture we begin to emphasize the importance of the oral presentation, and steps to be taken to succeed. This lecture marks the end of the first quarter of the lecture notes, and students need the same amount of time to present their mid-course presentation.

Introduction

It should be brought to the students' attention that the chart under "Essential thoughts in this lecture," that is described in each issue of this lecture-note series is there in order that the students incorporate similar charts in their future reports. In our experience, we also need to remind students to look ahead and do their work in a manner that this is their last and final presentation. In other words, they should expect to prepare several drafts of their final report before submitting.

Steps Towards Oral Presentations – Phase I

Students are reminded that soon they are required to give a brief presentation of their project. To that end the following list of items, which must be incorporated in their presentations should be read in class. But first, let us assume that we have an audience for these presentations, which consists of department faculty and other students in this class (or possibly visitors from other classes). Also, in this audience we have a group of outside visitors, who are members of technical staff (MTS) of the company that sponsors this project. These are as follows.

(1) engineers at all levels (from the manager to the field engineer),
(2) staff from purchasing and finance,
(3) staff from research department with advanced degrees, and
(4) Director, who can veto the entire project.

The notion of talking in front of all these people may make you nervous, but the whole idea behind this exercise is to help you overcome your fear and improve your oral communicational skills. We are here to help you with issues related to your project, which you may not see yourself, no matter how hard you work. This is not a contest, but rather the first step toward getting you ready for your final presentation.

Here are some general rules of thumb, which you should incorporate in your presentation. Incidentally, a major part of your time at your first few jobs (on average for your generation you should expect to change at least six jobs) is to spend on these seemingly trivial presentations over and over again. People in industry are obsessed about these presentations particularly. Your presentation must include in a glossary format the following slides or transparencies. In the following we use the word slide as an example, you can have transparency for your actual presentation.

Do not forget to talk about lessons learned as applicable in each entry below.

(1) Title of the Project, names of all participants, and the date of presentation, all must appear in the first slide. Clearly, you need to introduce everyone before moving to the next slide.

(2) Brief Introduction, entails all elements of your talk briefly. Do not use acronyms, unless defined right away. Also, start with something that is a common knowledge (relatively speaking) and build on that. For instance, in this slide you may talk about the following items, but you must go from one topic to another very smoothly. Give an explanation of each entry below.

(a) Problem statement.
(b) The solution selected, if more than one exists, and why this particular solution is chosen?
(c) The selected technical approach (just a name).

Here, you may also use footnote(s) to include your citations, for instance, when you mention the technical approach that you have selected, you need to mention the corresponding source in our standard for citations.

(3) Technical Approach, entails details of your contributions, and how your team have come up with this approach individually as well as collectively. Here is the place that you state your goals and plans in this project. One, or possibly two slides should include all your pertinent oral presentation of this matter (technical approach), mentioning the following as you deem necessary:

(a) Block diagrams of tasks.
(b) Personnel assignments and schedules.
(c) Finally, the test plan and demonstration as you see now.

(4) Budget, that entails all aspect of securing resources and funds and using them wisely. For instance, you may talk about resources that you are expecting from your client, if any. Also, in your slide include discussion of the following items.

(a) Cost breakdown to date.
(b) Anticipated future cost and how it would be secured.

This is the first phase of your oral presentation that you should begin to think about. We improve and revise the above steps, as new topics such as finance, marketing, and ethics are included in our future lecture notes.

Final Thought

As we approach the end of this semester or quarter, we need to collect our thoughts and get as much work done as we possibly can, in order to get to the actual design and testing. Thus, while we are breaking these procedures to you and hoping that you get ready one step at the time for the final presentation of this course, we expect that you do not lose sight of the bigger picture.

Closure

We remind students to consult with some of the students' samples work in Part Three to get ideas for preparing their slides.

Essential thoughts in this lecture

Issues.	Applicability to your project, if any.
Steps to prepare for the oral presentation.	Obvious!
Do you want to add anything else?	Please elaborate.

LECTURE SEVEN: *Technical Writing – Phase I*

Remark: In this lecture we begin to encourage students to pay more attention to their writing than before. Many of them are under the illusion that in their jobs they are only responsible for the technical accuracy and somehow someone else is writing the documentation. To educate engineering students to write is an enviable task.

Introduction

In this lecture we go over a few items, which were noted in your submitted reports, in order to clarify some of the essentials of technical writing and elements of style. One lecture is not enough to go over everyone's paper, but the following samples are indicative of the issues that we must be concerned with. In lecture five, we *pointed* out (why past tense, and when can we use present tense?) how difficult it is to compartmentalize one's thought and give proper steps to break a presentation or writing into segments, but we need to start somewhere and this place is a good beginning. Before we close this section, let us answer the preceding question. If we are referring to the ***event*** of presenting lecture five, *per se,* then we use past tense, because there is an element of time involved here, which was two lectures ago. But if we are referring to the *location* of a piece of information, which is now in a place called "Lecture Note Five," then we use present tense, because these notes are written completely and are available in front of us. Just as any textbook, which is written in certain order and as far as we are concerned we only want to know where a piece of information can be located. Therefore, please ***no more we will or we did explain!*** Section number and all those addresses in a document are for locating information. This explanation is due Professor T.J. Higgins.

Technical Writing – Phase I

Good writing begins with extensive reading. One cannot elucidate clearly if he/she does not read regularly literature, or technical papers in this case. Even experienced writers go through several iterations before putting out their final work, which often, and before the ink is dried, they wish they had used different words or different sentences. Therefore, be prepared to spend a major part of your time on writing and related documentation of your project. In many universities, there are writing laboratories, which help students to improve their skills, but we are pressed for time. There are several other ways that you can get help. However, be aware that some of these helps may actually handicap you, unless you are selective in your choices and insert your own inputs into the final draft of whatever document you are preparing.

Nowadays, for instance, most personal computers come with scores of designed forms and letters that all you need to do is to fill the blank and use these forms to communicate. But that writing does not reflect your own individuality and uniqueness. Your voice does not appear in these forms. Just imagine, you receive 100 resumes to review and all applicants have used the same format to express themselves, among many readily available. On the other hand, if only a handful of these applicants have used their own initiatives and/or have prepared their resume differently, then that small group will stand out in the crowd. That is the way we human being are programmed to react. We are sensitive to identifying these differences very quickly. Here, we are not trying to discourage you from using special forms available in your computers for some of your writing, but rather we are suggesting that be aware of some handicaps that may result when extensively and solely relying on pre-packaged forms in your computer. Remember that these forms are available to everyone. There are of course, many other capabilities in your computers, which enhance your writing and it would be a grave mistake not to benefit from that. For instance,

spell checking, which is readily available in most computers. It would be totally unacceptable to submit a paper with even one misspelled word. It would be considered an insult to do that! In some sophisticated computers there are also ways to check

grammar, but there again you must be careful not to become a slave to someone else's style. Be particularly careful about using words with

right sound, but wrong meaning. For example, "the computer is done," versus "the computer is down." Be particularly concerned about words with

multiple meanings. Avoid using such words altogether if you can, but if you cannot, make every effort to explain your intended meaning. Pay particular attention to

redundant words and phrases. You cannot detect these imperfections in your own writing no matter how hard you try. That is why it is absolutely necessary for you to let someone else reads your report. In fact, we suggest that you exchange proof reading tasks among your peers as a service to each other. Make sure that all sentences are

complete, and not dangling. Having said all the above, we now go back to the main premise of this note. Suppose, we know all the above, then how do we write a technical paper? The fundamental answer is we must first divide the work, which we are trying to write about, into several segments, and pedagogically connect those segments together as follows.

There is always a beginning to the work, and there is an end, and there are segments connecting these two. Each segment should be in at least one paragraph, or a section consisting of several paragraphs. It is always required to number each segment in any technical writing. Furthermore, if a segment is divided into smaller portions, then each portion has its own heading, which normally is under the main heading of that segment. For instance, the "Introduction," of a report is its Section 1, and is shown as:

1. Introduction.

If this segment (note that we still are in the same paragraph, and the reason that we have indented the above line is for the sake of demonstration) has other main elements such as the problem statement, *per se*, then that becomes its Section 1.1, 1.2, *etc.*. (Note that we use a double-dot here. One dot for the last word, *etc.*, and the other dot for the end of the sentence, but in many contemporary style only one dot is used.) Thus, we have as, an example,

1.1. Problem Statement. (Note that we also use 1 dot 1 dot.)

In some cases those portions or sub-sections are labeled by A, B, *etc.*. In some cases no dot is used at the end of the section heading – many put the dot as shown above.

It is always easier to read a report if it has been coherently sub-divided into sections, each has its own element and/or contribution to the overall body of the report. Each report closes with some concluding remarks or conclusion, where an assessment of the accomplishment or the lack of it, as set out in the introduction, is made by the author. Here, the author may also state his/her future goals based on these accomplishments thus far.

All references must be collected and numbered in alphabetic order (that is our preference and we will elaborate on that in class) at the end of the report. Please use our citation style for reporting these references. Using footnotes are allowed throughout the report, but be aware that these can become very distracting and should be avoided as much as possible. In any event, when using footnotes, number them sequentially throughout the report for possible future reference.

Going back to our earlier comments in this course, when all are said and done, more that 90% of our writing task is about how successfully we can break the actual work or experiment into identifiable segments, or pieces. This segmentation is used in order to describe the process of putting back this construction together in a step by step process. Then writing about this experience is a matter of explaining how each segment is defined and what is its main function.

These comments are just the tip of the iceberg, and we have so much more to talk about, but we hope to go over more of this type discussion in our future lectures.

Finally, you may wonder how can someone become a very good technical writer? Considering that English is not the native language of most engineers or engineering students. The short answer is technical writing for a task that is compartmentalized has very little to do with English, *per se,* and can be managed by every engineer. But the final answer on how to become a good technical writer is three words, practice, practice, practice!

Final Thought

The single most important thing that you must remember about good technical writing is ***consistency***. There is nothing more distracting and noticeable about a poor writing than its inconsistency.

Closure

The class ends with reading some more samples of written work of students and instructor as well as other related matters.

Essential thoughts in this lecture

Issues.	Applicability to your project, if any.
Steps to prepare a technical writing.	Obvious!
Do you want to add anything else?	Please elaborate.

Question: Can you show us one sentence with all different tenses?
Answer: Last summer, I ***studied*** [past event] this topic and ***found*** [past event] scores of authoritative articles, which some ***have been read*** [present event], but many ***will be scrutinized*** [future event] next summer, meanwhile, my research on this matter to date ***is reported*** [location, always present] in Section 3.1.

LECTURE EIGHT: *Marketing High-Tech Products*

Remark: To expose engineering students to highly specialized topics in business and management, we have designed four lectures (Eight, Nine, Eighteen, and Nineteen) describing those topics. In this regard we have been fortunate to know Mr. Kayvan Mozaffari – an MBA from University of Southern California, and a former M.Sc. student of this instructor at the State University of New York at Stony Brook. He has developed and has delivered these lectures with numerous practical and proprietary examples at the University of California at Riverside. We hope that exposure to these highly specialized topics will inflame the thirst in many of our engineering students to pursue studying the business aspects of the engineering profession.

The purpose of this lecture is to introduce engineering students to the marketing disciplines. History has shown that business and marketing aspects of technology are as important to the success of high-tech products as a robust engineering design and reliable manufacturing process. The marketing of high-tech products is a specialized topic that requires an in-depth study of this subject matter. Students with an interest in marketing of high-tech products are encouraged to read the following references in addition to taking courses in marketing with an emphasis in product marketing.

References:

[1] W. H. Davidow, *Marketing High Technology: An Insider's View.* New York: The Free Press, 1986.
[2] B. Heldey, "Strategy and the business portfolio," *Long Range Planning,* vol. 10, no.1, pp. 9-15, February 1977.
[3] P. Kotler, *Marketing Management: Analysis, Planning, Implementation, and Control.* Englewood Cliffs, New Jersey: Prentice-Hall, 6th Edition, 1988.
[4] G. A. Moore, *Crossing the Chasm: Marketing and Selling Technology Products to Mainstream Customers.* New York: Harper Collins Publishers, 1991.
[5] _____, *Inside the Tornado: Marketing Strategies from Silicon Valley's Cutting Edge.* New York: Harper Collins Publishers, 1995.

Introduction

Topics to be covered include basic marketing definitions, concepts, fundamentals and disciplines; as well as product marketing for actual high-tech products. These topics are presented in this and the subsequent lecture note.

In today's business climate, most companies have little or no competitive advantage over either large/global and/or small/local competitors. No longer, anyone of these companies has any significant and measurable lead on quality, creativity, delivery, and capital over the rest. Therefore, marketing plays a very important role in creating a major differentiation between products.

I. Basic Definitions [3]:

The following core marketing definitions are from Kotler's book [3]. This book is a recommended reading for students with an interest in marketing management.

Marketing: Marketing is the process of creating and exchanging products and services between individuals and organizations to satisfy their *needs* and *wants*.

The core concept of marketing is shown in Fig. 1. It begins with determining customers' needs and wants to create products and services with a unique set of values for these customers. Marketers help with selling these products and services in the marketplace.

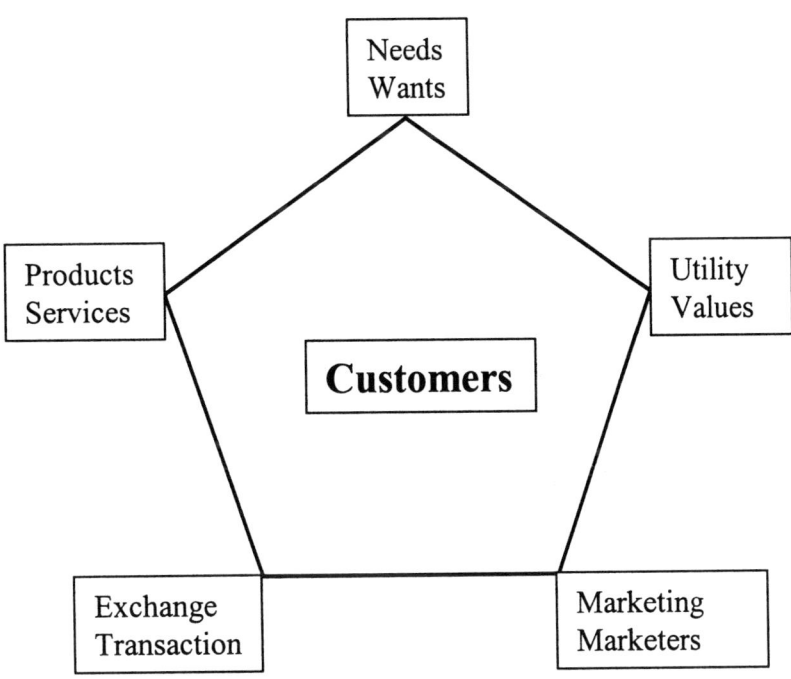

Fig. 1. Marketing core concepts [3].

Needs and Wants: Satisfying human needs and wants are the foundation for the marketing discipline.
- **Needs** – A human state of felt deprivation requiring basic satisfaction for survival such as food and shelter.
- **Wants** – This refers to a human desire for a particular satisfier to fulfill his or her needs. For instance, needing food and wanting something specific (a particular brand of hotdog) is a relevant example.

Products: Products and services are offered to satisfy needs and wants of people.

Utility and Value: Product selection is made at this level. To satisfy a need, a consumer will choose a specific product based on its value and utility.
- **Utility** – An estimate of product potential to satisfy one's needs.
- **Value** – Amount of utility gained per dollar to satisfy one's needs.

Exchange and Transactions: Marketing emerges as a result of people exchanging products and services to satisfy their needs and wants.
- **Exchange** – An act of obtaining a product or service and in return providing a product, service or money.
- **Transactions** – When two parties agree to exchange and trade values.

Markets: A set of potential customers with a common set of particular needs and wants with purchasing power for particular products and services.

Marketers: Marketers actively seek exchange of products and services with others to satisfy human needs and wants.

A simple marketing structure that consists of sellers and buyers is shown in Fig. 2. In this simple marketing structure, buyers receive products and communications (about products) from sellers, and in exchange, buyers pay money for products and information (regarding their needs and wants.)

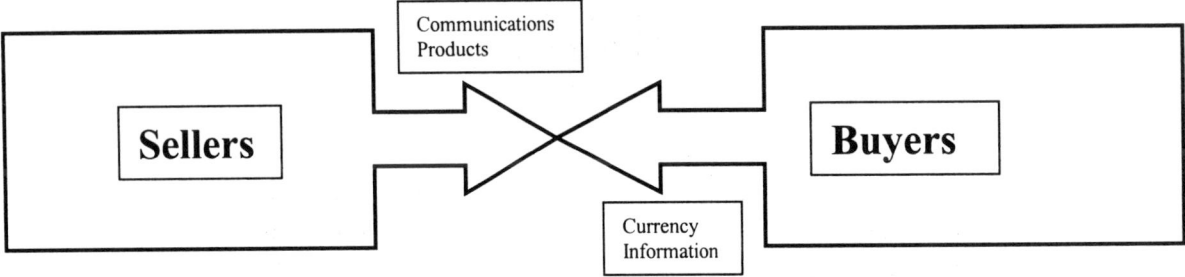

Fig. 2. Marketing structure [3].

II. Market-Oriented Strategic Planning [3]

Strategic planning is the managerial process of developing and maintaining a viable fit between the organization's objectives and its changing market opportunities. The aim of strategic planning is to shape and reshape the company's business and products so that they combine to produce satisfactory profit and growth.

As shown in Fig. 3, the strategic planning process begins with corporate planning at the top executive level, followed by business and product lines and services planning. This strategy followed by execution and controlling processes, are practical aspects of the overall planning process.

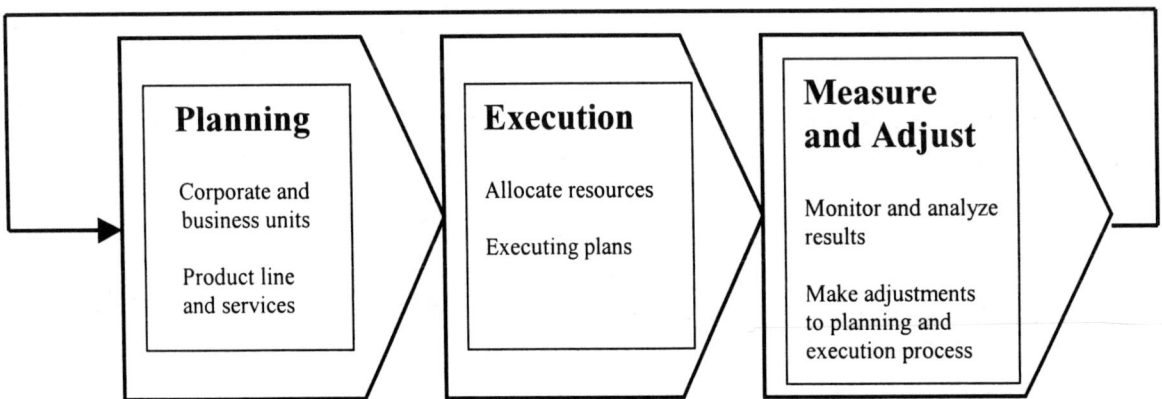

Fig. 3. The feedback process for strategic planning, execution, and measure and adjust [3].

Corporate planning process includes defining the mission and strategic goals for business units, analyzing markets and evaluating products that the corporation owns, and identifying new markets to enter as depicted in Fig. 4. The purpose of this process is to ask a series of fundamental questions and find answers that keep organizations focused. Ultimately, the benefit is to enhance the overall business growth measured by an increase in profit and sales.

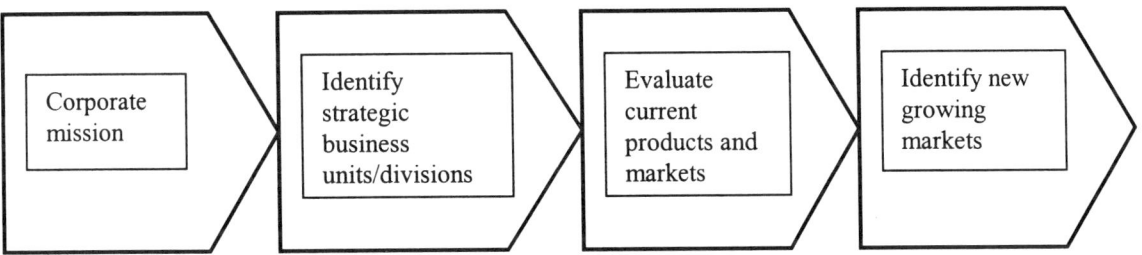

Fig. 4. The process for the corporate planning [3].

The strategic planning process for each business unit is shown in the Fig. 5. This figure consists of defining a focused mission, analysis of the internal and external environment (the internal strengths and weaknesses, and the external opportunities and threats – also known as SOWT). Also, this figure shows goal setting, defining focused strategies regarding how to achieve these goals, as well as developing programs for carrying out strategies, allocating resources, executing programs and strategies. Finally, this figure shows the need of implementing a feedback system to continually monitor and make necessary adjustments to control its progress.

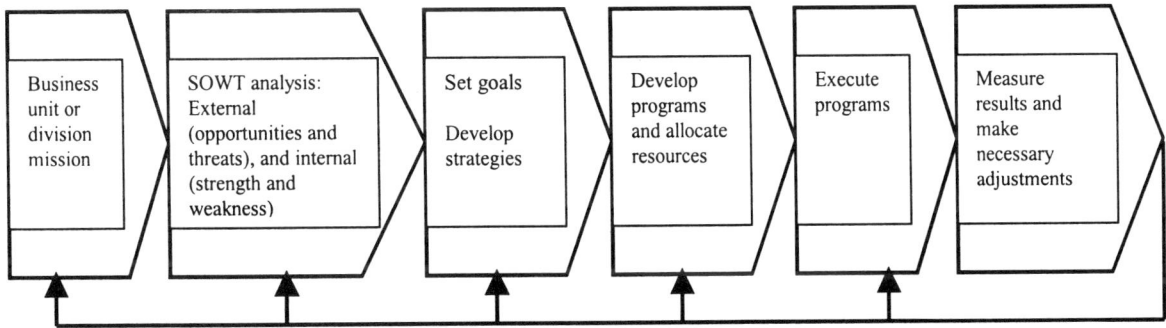

Fig. 5. The feedback process for the business unit/division strategic planning [3].

III. Growth Strategies [2] & [3]

There are many strategies to enhance the growth of any company. Among those, three of the most significant ones are:
- *Intensive growth*
- *Integrative growth*
- *Diversification growth*.

The ***intensive growth***, which is the growth within current business, is as follows. The high-tech marketing strategies for most small and medium size companies usually revolve around creating intensive growth opportunities through market penetrations, market development, and product development. The ***integrative growth***, however, is defined as expansion through building or acquiring related businesses to the existing businesses. Integrative growth strategy can be *backward* (buying companies that supply parts), or *forward* (buying distributing channels such as wholesalers and retailers), or *horizontally* (buying out competitors). ***Diversification growth*** strategy, on the other hand, is the basis for creating conglomerate companies, which is defined as adding businesses that are attractive, regardless of their relevance to the current business of the initial company.

The famous Boston Consulting Group's Growth-Share Matrix, as shown in Fig. 6, is a great tool by which we analyze companies, business units, or divisions and their market share and growth potentials. This matrix shows the relative position of business units, or divisions, or product lines (owned by a company) according to their market growth rate and their relative market share. According to this matrix, there are four categories that a business unit, or a division, or a product line can be characterized under.

Star Category: • High Growth Rate • High Market Share	*Question Mark Category:* • High Growth Rate • Low Market Share
Cash Cow Category: • Low Growth Rate • High Market Share	*Dog Category:* • Low Growth Rate • Low Market Share

Fig. 6. The Boston consulting group's growth share matrix [2].

1. ***Star Category*** – Units with high market growth rate and high relative market share
2. ***Cash Cow Category*** – Units with low market growth rate and high relative market share
3. ***Dog Category*** – Units with low market growth rate and low relative market share
4. ***Question Mark Category*** – Units with high market growth rate and low relative market share.

The result of this categorization is used to pursue an alternative strategy for each of the corresponding business units, or divisions, or product lines based on the company's overall objectives. These alternatives include building, holding, harvesting, or divesting. For example, when analyzing product lines, a likely conclusion would be to hold strong product lines in the Cash Cow Category, and build product lines in the Star Category and high performing Question Mark Category. Harvest product lines in the Dog Category, low performing Question Mark Category, and weak Cash Cow Category. Divest from product lines in the Dog Category and Question Mark Category, which drag down the overall company profit and have no strategic benefit.

There are many tools by which each business unit, division, or product line is rated in terms of its market attractiveness and competitive positioning. The important point from this analysis is to pursue attractive markets when the business unit, or division, or product line has the competitive advantage and strength to succeed in those markets. Again, to have a good chance to achieve great results, both conditions of attractive markets and a strong company with competitive advantage must exit. The above tool is used in the management of the company to evaluate and communicate objectives and strategies.

IV. Marketing Management Process and Marketing Plan [3]

The marketing plan follows the marketing concept and business strategic plan. A marketing plan is narrowly focused on specific market(s) and product(s). It describes the necessary details for marketing strategies and tactics to achieve marketing objectives for products. The marketing plan is the single document that is used to communicate and coordinate all marketing activities.

The content of a marketing plan, which is outlined below, is the most important outcome of the marketing management process. This general outline should be used as a guideline to develop the marketing plan and should be modified according to specifics of product/market objectives.

Table 1. A general marketing plan outline.

1 – Executive Summary	
2 – Marketing Summary	2.1 Category
	2.2 Market Segmentation
	2.3 Market Situation
	2.4 The Customer
	2.5 Competitors
	2.6 Market Share
	2.7 Distribution Situation
	2.8 The Macro-Environment
3 – Assumptions and Unavailable Information	
4 – Opportunities and Issues Analysis	4.1 Strengths, Weaknesses, Opportunities, and Threats (SWOT) Analysis
	4.2 Issues Analysis
5 – Evaluation of Alternatives	
6 – Objectives	6.1 Financial Objectives
	6.2 Marketing Objectives
7 – Marketing Strategy	7.1 Unique Selling Proposition
	7.2 Target Markets
	7.3 Positioning
	7.4 Product (Life Cycle, Benefits and Features, Release Formats, Packaging)
	7.5 Marketing Channels (Retail, Direct, Value Added Reseller/Systems Integrators, Online)
	7.6 Pricing Strategy
	7.7 Promotion (Advertising, Sales Promotion, Sales Force, Direct Mail, Public Relations, Trade Shows)
	7.8 Engineering Resources
	7.9 Marketing Research
8 – Implementation	8.1 Launch Date
	8.2 Action Plan and Schedule
9 – Control	
10 – Financial Statements	

An important task for the marketing manager is to refine the marketing strategy and to set marketing objectives in accordance with the company's business objectives. When the marketing budget (usually as a percentage of sales) is set, the marketing manager must decide how to allocate resources among various elements of the marketing mix. The elements of the marketing mix are called the Four **P's**: ***product, price, place,*** and ***promotion***. These four **P's** of the marketing mix and some of their variables are shown in Fig. 7.

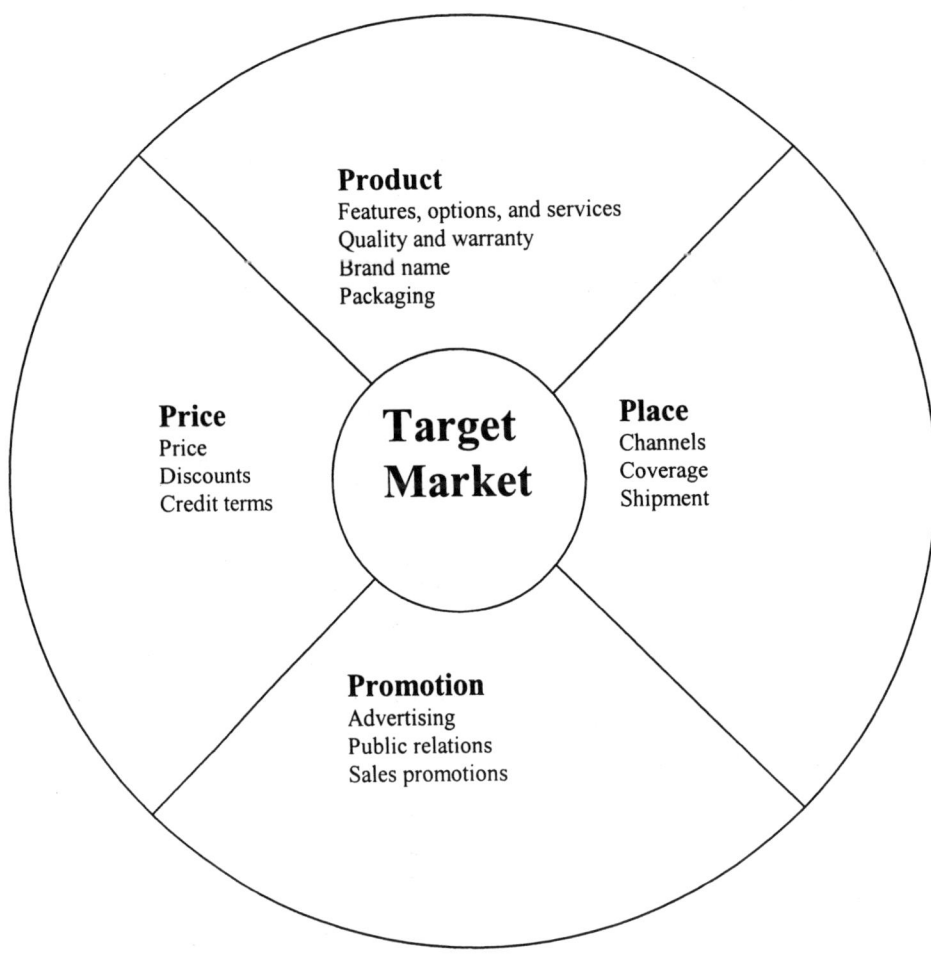

Fig. 7. The four P's (Product, Price, Promotion, Place) of the marketing mix [3].

The marketing mix strategy as illustrated in Fig. 8, is designed to reach both the distribution channels and the end-user customers. The marketing manager must prepare two programs, namely, the offer mix (products and price), as well as the promotion mix (sales promotions and campaigns, advertising, and public relations).

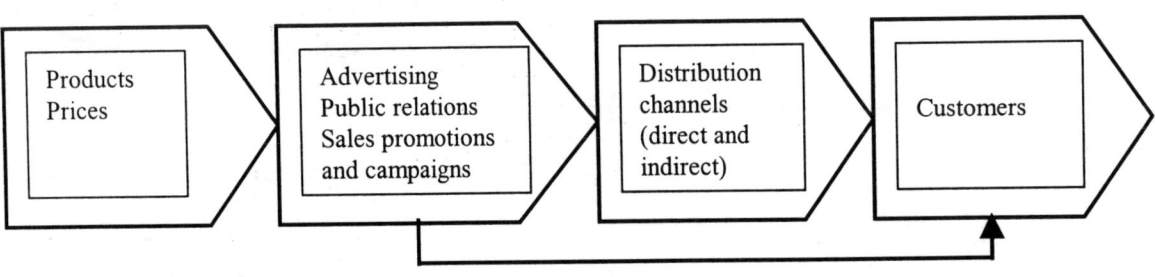

Fig. 8. Strategy for the marketing mix [3].

Final Thought

The fundamentals of marketing theory are the basis for high-tech product marketing strategy. Understanding needs and wants of customers is as important to the success of high-tech products as making robust technical designs.

This thought goes back to the first idea in this course, namely, no one should spend time and money on any project that has no future. To be successful in your career you need to evaluate these issues continually. This is why we have sponsored these specialized lectures on management and business aspects of an engineering project. In order that you will be fully alert of these issues, which are not peripheral but rather substantially related matters to the main issue in hand – making profit by selling a product and thus sustaining an engineering enterprise. This notion to sustain the enterprise is of paramount importance and is the only prerequisite to executing the social and philanthropic responsibilities of the company, but first thing first. Do not lose sight of that.

Closure

Here, we have discussed marketing definitions followed by marketing management and strategic planning processes. The most important outcome of strategic planning and marketing management process is the development of the marketing plan. Any marketing plan communicates marketing objectives and coordinates marketing activities with all elements of the company.

Essential thoughts in this lecture

Issues.	Applicability to your project, if any.
Why is understanding of the high-tech marketing important?	It gives you a top view of where you and your work fit in the product value chain.
Do you want to add anything else?	Please elaborate.

LECTURE NINE: Marketing High-Tech Products (Continued)

Remark: The following items are the continuation of the preceding lecture by our guest speaker Mr. Kayvan Mozaffari – a former student of this instructor.

Introduction

The purpose of this lecture is to introduce engineering students to the high-tech product marketing concepts, which is a very specialized topic. Students with interest in this subject are encouraged to consult the previous list of references. This part focuses on product marketing for high-tech products.

I. Product Marketing and Product Life Cycle (PLC) [3]

Both products and markets follow a demand life cycle curve, as shown in Fig. 1, which passes through the six stages of *emergence, fast growth, slow growth, stabilization/maturity, decline,* and *end of life*. Products have limited life, and their contributing profits change based on their life cycle stage, which requires employing different marketing strategies at different stages.

A new market emerges when an innovator creates a new product that meets an unmet demand. Upon product success, demand for the product will increase which in turn brings competition to the corresponding market. When the market matures, the demand for the product stabilizes and its growth rate slows down, but now there are more competitors in the market place than before. Because of the increased competition and incremental innovations, the market segments go through a period of *fragmentation* and *consolidation*. The next stage brings a decline in demand for the current technology/product, which is usually the result of a new and better technology/product with higher utility than that in the preceding stage. Finally, the last stage of the product life cycle is the end-of-life stage for the current product.

The PLC stages provide a tool for product marketing managers to develop marketing strategies and implementation plans, as well as organizing appropriate cross-functional teams to execute these plans. This strategy improves the success rate of a product and increases profits in a competitive marketplace. However, executing a marketing plan requires allocating necessary budget, which is primarily based on the expected profit potential, which brings us back to identifying accurately the PLC stage that the product is going through.

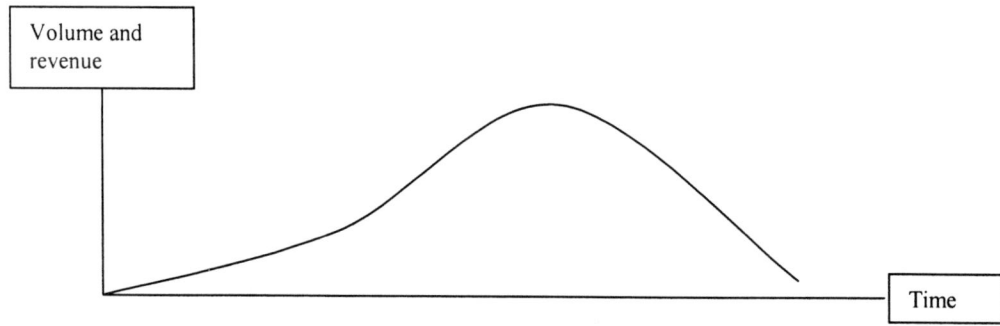

Fig. 1. Product life cycle [3].

II. High-Tech Marketing [4] & [5]

References [4] and [5], are excellent starting points for anyone interested to learn high-tech marketing strategies. These books are recommended readings, especially for those with an interest in marketing and having an entrepreneurial aspiration.

According to Moore, high-tech market development follows the *technology adoption life cycle*, which is described in more detail in Section III. This concept is introduced and explained in great detail in [4]. Moore defines the markets for high-tech as: customers with common needs for products, but when they are ready to buy, they reference each other [4]. Therefore, building a list of influential reference customers is the most important step in a successful high-tech marketing strategy.

III. Technology Adoption Life Cycle [4] & [5]

The technology adoption life cycle is illustrated as a bell curve and is shown in Fig. 2. According to Moore, the noticeable segments or stages are *Innovators (I), Early Adopters (EA), Early Majority (EM), Late Majority (LM),* and *Laggards (L).* The mainstream market, which consists of Early Majority and Late Majority make up about two thirds of the buyers, and the rest make up the remaining one third of the buyers. The technology adoption life cycle is a significant model, because it helps high-tech marketers to communicate the appropriate message to the right audience.

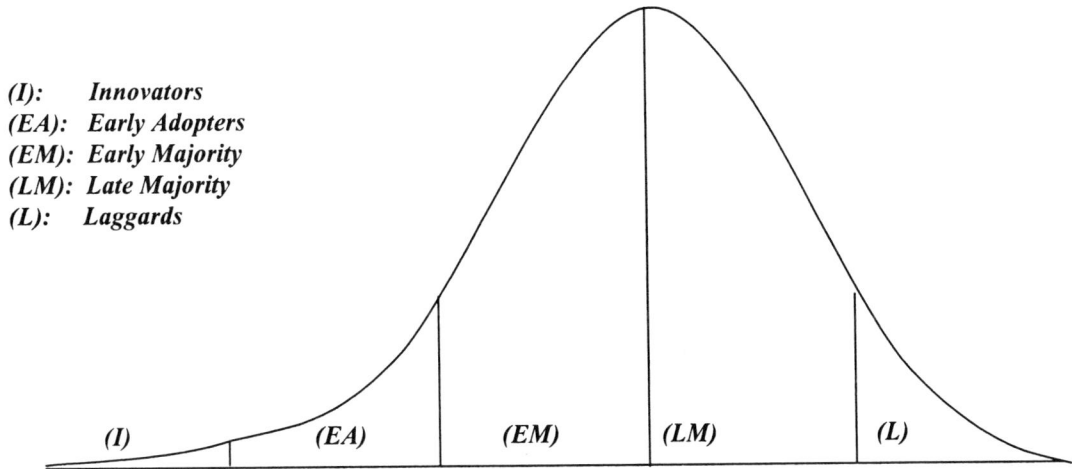

Fig. 2. The technology adoption life cycle [4].

The following list briefly describes each of the five segments in the technology adoption life cycle.

- ***Innovators:*** They are "techies," who usually buy a brand new technology and product just to find out about its capability. They normally do not control the major spending budget, but they do influence people who do, and without their endorsement a new technology will not have a chance to succeed.

- ***Early Adopters:*** They are truly the first powerful buyers with real money to spend. Their motivation is to bring a real competitive advantage over the existing norm in their industry. They validate the product and technology, and legitimize its real potential for the mass market.

- ***Early Majority:*** They are neutral to new technology but they usually are in charge of infrastructure and mission-critical systems. They believe in evolutionary changes and must see a proven track record and strong reference from users of a new product and technology before deploying it. These customers as a group buy from the market leader to ensure its survival, which results in creating a rapid growth and profitability for the producing firm.

- ***Late Majority:*** They are very price sensitive and demanding. They are not very comfortable with technology products. They only buy technology after standards are fully established (when no more R&D is required). They usually buy from major leading vendors.

- ***Laggards:*** They do not want technology. The only technology this group gets is already imbedded in another product that they buy. They make up a small portion of the buyers.

The above characterization is very important for high-tech marketers. It helps marketers to create marketing programs and allocate resources in accordance with what particular marketing stage the technology and product is going through.

IV. The Chasm [4] & [5]

As shown in Fig. 3, there is a gap between the Early Adopters and the Early Majority. Moore calls this gap: ***"The Chasm",*** which he has discussed in great detail in [4].

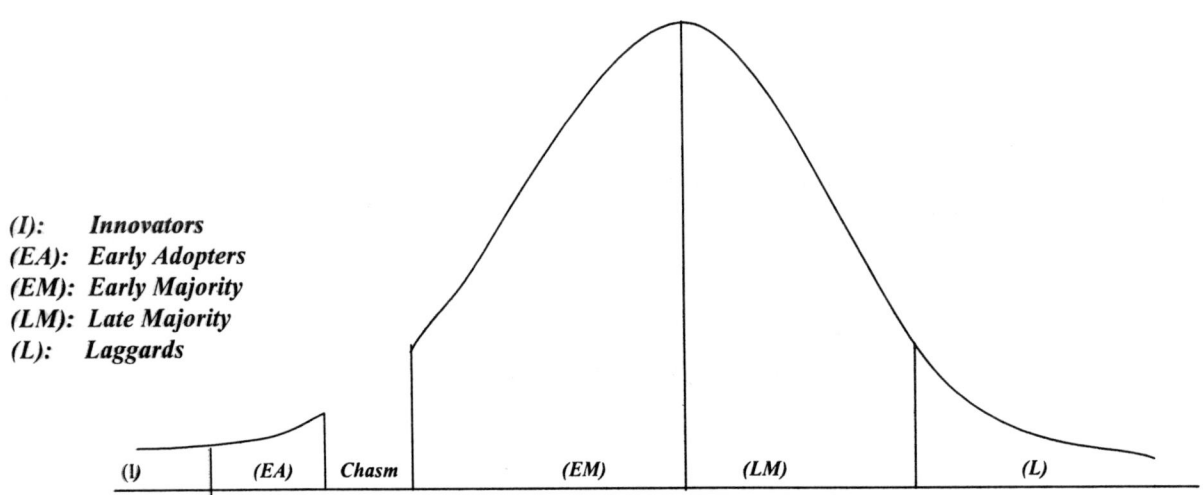

Fig. 3. The Chasm concept [4].

The idea of the Chasm in the high-tech market refers to a real disruptive innovation that gets initial acceptance by the early market, which according to Moore, consists of Innovators and Early Adopters. But then for a time period, sales will slow down, which signals that the product is in the Chasm period. If the product successfully crosses through this Chasm period and gains the acceptance of Early Majority customers, then the product has entered the mainstream market, which according to Moore, consist of Early Majority and Late Majority customers. The mainstream customers make up about two thirds of all the buyers and generate a large amount of revenue. Yet crossing the Chasm is not easy and many new products may not actually make this transforming journey safely.

The Early Majority market is subdivided into *small niche markets* and *large vertical markets* in its evolutionary process relative to mainstream markets. Niche and vertical market are one type of the end user customer segment. An example of a vertical market is the end user customers in the aerospace industry. We call the niche market as a specialized portion of a general market segment. Finding many small niche markets for a new technology and product is the first significant sign that it has successfully crossed the Chasm. As the adoption rate for the technology increases, a few large vertical markets will develop, which indicates the new technology is maturing and is ready to enter the mainstream market.

V. Disruptive Innovations [4] & [5]

The Technology Life Cycle shows how disruptive innovations get adopted among five buying group segments (i.e., Innovators, Early Adopters, Early Majority, Late Majority, and Laggards as shown in Fig. 2.) Disruptive innovations are revolutionary and need to be adopted by buying groups, which promote new markets and power structure. This type of interruption threatens the existing power structure. The established power structure is more in favor of gradual change in order to maintain its status and existence. For example, when the mainstream market adopted personal computers, a new power structure replaced the established power structure by mainframe computer companies.

VI. Whole Product Planning [1], [4] & [5]

Understanding the whole product concept is very important for product positioning and perceptions, and therefore its success. As shown in Fig. 4, the whole product concept consists of two parts. The first part is the generic product, and the additional parts are the peripheries (which are additional products and services, that make up the whole product). The whole product is what customers expect to receive. You can find more about the whole product concept in [1].

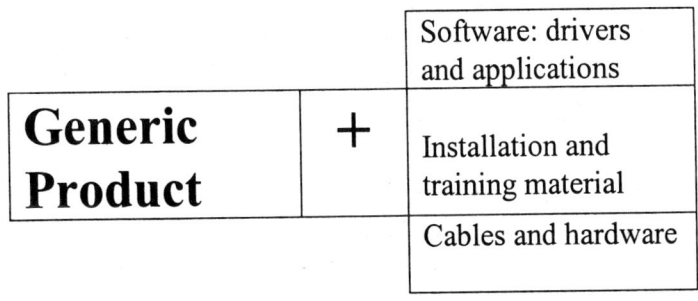

Fig. 4. The whole product concept [4].

VII. The Value Proposition [4]

The value proposition, as shown in Fig. 5, consists of three parts: customer, application, and product. To formulate the value proposition and a compelling reason for the customer to buy in a target market, high-tech marketers must address all three parts. Realistically speaking, the product is very difficult to change, so the remaining two elements of customer and applications are used to come up with the value proposition for a target market. The process of finding the right value proposition for the right target market will help us to examine various target markets and to select a market based on the most compelling reason to buy.

For example, in order to come up with a compelling reason to buy, in a value proposition statement for the mainstream target market, the value proposition statement must have:
- Gain dramatic competitive advantage that improves the operation.
- Significantly improve productivity in a mission critical area.
- Dramatically reduce operating costs.

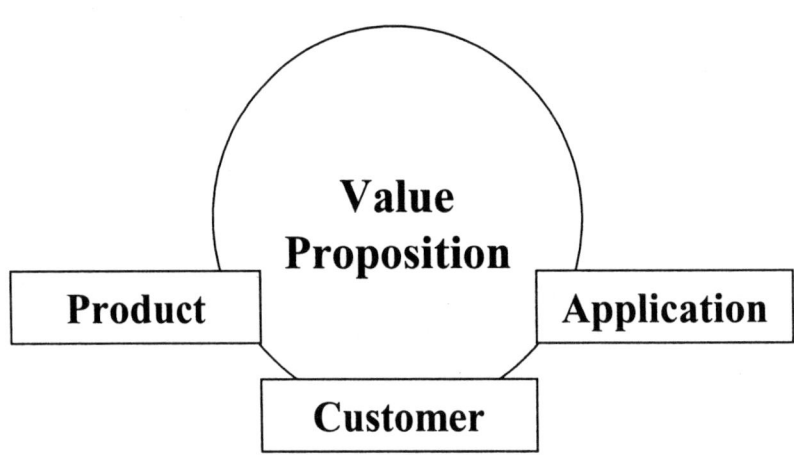

Fig. 5. The value proposition [4].

VIII. Competitive Positioning [4]

Positioning is the most important and least understood concept in the marketing of high-tech products. The most important goal of positioning is to make it easy for the target customer to buy your product. Moore describes the positioning goal as creating and occupying a space in your target customers' head for the best buy of a particular product category. The positioning is dynamic and changes as products go through various stages of the technology adoption life cycle.

The positioning process consists of four elements [4]:
- Making a claim is a fundamental part of the positioning statement.
- Showing evidence that substantiates this claim is the next important part.
- Communicating the proof of this claim to the right audience, and
- Controlling the process by anticipating competitor's moves and the ability to counter them are the last elements of a positioning process.

Final Thought

A deep understanding of high-tech marketing issues will enhance one's ability to invest resources judiciously while developing a new product. The goal is to deliver products in accordance with the current and future market demands. The technology adoption life cycle is a great tool to achieve a conceptual understanding of various stages of high-tech market demands and should be utilized regularly in any business planning.

It is also to be noted that, if you are not working very hard to make your product obsolete, then someone else is trying very hard to that for you. Do not stop thinking and searching for new ideas. In fact, the best and the most productive time to come up with a new product is when you are enjoying the comfort zone created by your last idea.

Closure

Here, we have described the technology adoption life cycle as an important marketing tool for high-tech marketers to develop marketing programs. Also, we have briefly discussed some very important concepts in high-tech marketing such as disruptive innovations, standards, whole product planning, value proposition, and competitive positioning.

This and the preceding lecture have been delivered sequentially by our guest speaker who has also shown a few practical and proprietary case studies to motivate students.

Essential thoughts in this lecture

Issues.	Applicability to your project, if any.
Why is the understanding of high-tech product marketing in the context of product life cycle important?	Your professional success will often depend on getting involved with projects that are in the high-growth markets.
Do you want to add anything else?	Please elaborate.

LECTURE TEN: *In Preparation for the Design Review*

Remark: As we approach the middle of the course, students should be reminded of the next goal, which in this case is the mid-course design review. In this lecture we concentrate on that and describe the key issues.

Introduction

We are about to finish the term, and soon you will be asked to give a mid-course presentation of your project. To that end, we suggest that Lecture Note Six to be reviewed again. As a reminder, use no more than five slides or transparencies as described in that note to prepare your demonstration. Emphasize the block diagrams approach, which connect different parts of your project with different contributions of your team members. Although we help you with this preparation, you need to begin reviewing your slides with your technical faculty advisor as soon as those are ready.

Although it is very early to talk about the final report, these reviews and their corresponding documentation must be prepared with an eye on the final report. In that report you need to specify the purpose of your project by summarizing the design and the design process qualitatively and quantitatively, and convey the finished project to the client. Finally, you must provide a platform for future improvement and expansion of your project. Your documentation must conform to our format and style as described before, and will be refined subsequently.

It is imperative to understand the meaning of specifying the purpose of your project by summarizing the design objective and the design process qualitatively and quantitatively. We will be elaborating on these matters in class and you need to pay a particular attention to the thought behind this difference. Here, the course instructor should plan to explain these as they fit in the corresponding program. We have learned that many students do not know more than half way into the course, how to order parts for their design. Often they spend large sums of money to purchase the wrong parts because they have no clue about their design objectives.

Parts List

It is also that time of the term, at which you need to have your parts list and know all the projected hardware needed to complete the task. Many of you have already secured these items and/or know how to get those. For those of you who still are not sure where you may get these items, please consult with your technical faculty advisor as soon as possible. The department does not have specific budget for the cost associated with this course, nevertheless it is your responsibility and a major part of your final report requirement to have this list with the name of corresponding vendors available. Keep a record of all your receipts. Thus, be optimistic and assume that you may actually get paid!

Do not take off for the break, unless you are absolutely certain that when you return you have all the necessary parts to construct your projects.

Final Thought

This is the time to reflect and see how to improve our standing in the course. We have found that students are either hopeful about their projects and thus are extremely happy, or just the opposite. We ask students to share their feelings with us in order to help them to get on the right track. Stress and frustrations are part of the picture, but these should be managed for better results. Some students need help and we should be prepared to give them instructions.

Closure

The class ends with comments regarding our expectation to have 30-minute meeting with each and every group before the end of the term. We hope to assure that everyone is on the right path.

Essential thoughts in this lecture

Issues.	Applicability to your project, if any.
Each and every piece of these lecture notes forms a tile on the final mosaic of this course. Try to assemble those such that it portrays your contributions to the underlying project, and your learning of issues discussed in this course.	Obvious! However, you may claim that, for instance, your project has nothing to do with certain issues discussed in class. Clearly, and, if that is the case, then you must justify your claim.
Do you want to add anything else?	Please elaborate.

LECTURE ELEVEN: Ethics

Remark: In this lecture we emphasize the importance of having a code of ethics – in particular, for engineers like us, and we review the IEEE code of ethics as a model.

Introduction

Throughout history, great emphases have been placed upon a higher sense of duty and public service. Ethics and civic responsibilities as carried out by citizens are the cornerstone of any civilized society. Clearly, as the rest of the society becomes more aware of the impact of engineering and technology in their daily lives, they have more respect and appreciation for engineers than before. In many foreign countries engineers are on top of the list for professionals and are respected greatly.

In this country however, some people do not know much about the impacts of engineers on their lives, and they do not know that engineers have explicit codes of ethics to adhere to, and the violation of these rules may grounds for disciplinary action. Thus, we propose to post the following IEEE code of ethics in your office, in order to affirm your high standard. Similarly, for other engineering societies there are codes of ethics which should be displayed.

In this lecture we review the resolution of the IEEE Ethics Committee convened in 1990 and approved by the IEEE Board of Directors the same year. For more information please consult the IEEE Ethics Committee web site at http://www.ieee.org/committee.ethics.

IEEE Code of Ethics

"We, the members of the IEEE, in recognition of the importance of our technologies in affecting the quality of life throughout the world, and in accepting a personal obligation to our profession, its members and communities we serve, do hereby commit ourselves to conduct of the highest ethical and professional manner and agree:

(1) To accept responsibility in making engineering decisions consistent with the safety, health, and welfare of public, and to disclose promptly factors that might endanger the public or the environment.
(2) To avoid real or perceived conflicts of interest whenever possible, and to disclose them to affected parties when they do exist.
(3) To be honest and realistic in stating claims or estimates based on available data.
(4) To reject bribery in all its forms.
(5) To improve understanding of technology; its appropriate application, and potential consequences.
(6) To maintain and improve our technical competence and to undertake technological tasks for others only if qualified by training or experience, or after full disclosure of pertinent limitations.
(7) To seek, accept, and offer honest criticism of technical work, to acknowledge and correct errors, and to credit properly the contributions of others.
(8) To treat fairly all persons regardless of such factors as race, religion, gender, disability, age, or national origin.
(9) To avoid injuring others, their property, reputation, or employment by false or malicious action.
(10) To assist colleagues and co-workers in their professional development and to support them in following this code of ethics."

The Kernel of the Above Code

Reviewing the preceding summary reveals that we may break this list into the following categories in conjunction with each practicing engineer in any enterprise. These categories and the corresponding relationships with others are summarized in the following matrix.

Issue below versus:	Public	Clients	Colleagues
Practicing Engineer Issue above versus itself: Self-improvement Items (5), (6), and (7)	Safety (1) Bribery (4) Fairness (8) Respect (9)	Conflict of interest (2) Honesty (3) Disclosure (6)	Professional Development (10)

Discussion

In this lecture we discuss the implications of the preceding code and give examples to elucidate the subtleties of these issues. Each course instructor may come up with similar examples that are suitable for the corresponding program. Perhaps one should also be more explicit about the ethical obligations toward the profession as well.

Final Thought

We need to elaborate our understanding of these matters as pertains to this course and our overall training as engineers in order to have a productive career.

Closure

The class ends with the announcement regarding our future and the pending plans for the course.

Essential thoughts in this lecture

Issues.	Applicability to your project, if any.
IEEE Code of Ethics. Or, similarly the ASME or any other engineering society's codes of ethics.	Obvious! However, you may claim that your project has, for instance, no harmful environmental effect, then explain why? In other words, the burden of proof to uphold the above code is upon you, and you must make the case for yourself.
Do you want to add anything else?	Please elaborate.

LECTURE TWELVE: Design Review Preparation – The Final Thought

Remark: The purpose of this lecture is to review important rules for the mid-course presentations and prepare students for the event.

Introduction

We are almost half way to the end of this course and students should be instructed to undertake a few steps in preparation for their mid-course review. Particularly, they are required to complete the following assignments among perhaps several others, as the course instructor may deem necessary.

(1) Prepare, *in consultation with your technical faculty advisor,* your slides/transparencies, within our standard, and using *five pages* maximum. **Remember you have only 15 minutes for your presentation!**

(2) Write, *in consultation with your technical faculty advisor,* your **final** report within our standard *(as of today).* In other words, use the style and procedures, which are described in class as of today, assuming that this copy is your final report, which is in fact the first draft of that report. Please note well that this report is prepared in order to support your presentation. That means, during your presentation, if anyone asks for additional details about your project, then you may refer to your report, section such and such, for an answer. This report cannot be informal and should contain all information in your slides and more.

(3) Make necessary arrangement with your technical faculty advisor regarding the best time that he/she is available for your demonstration. All within our allotted times as announced in class. Make the appointments for at least **two** sessions. If your advisor is not in town during that period, then tell him/her to send us a memo in order that we can finish the job in his/her absence.

(4) Make one **paper** copy of the above (1) and (2), plus the result of item (3), then bring all of those to the office of the course instructor by the deadline announced in class.

We are pleased to acknowledge that all students generally comply with most parts of these instructions. Upon receiving the above information a schedule would be developed for our "Mid-Course Design Review," and will be executed during the announced period.

Before proceeding further, let us remind you one more time that it is expected your *technical presentations* convey the following attributes:

(a) A complete description of your design project with all its components.
(b) All experimental data corresponding to each project must be given and analyzed as best as possible.
(c) All elements of this course, as pertains to your project, must be explained in a manner that the audience accepts your project as a genuine design project meriting your registered credits.

Advisory Evaluation

Students are informed that faculty members, who will sit in their presentations, are asked to comment on students' performance, as a guideline to them. In particular, the faculty members are asked to rank each group on a scale of 1 to 10 on the following items.

(1) The task's organization. (*Comment:* Please, pay particular attention to the contribution of each member. Unless stated explicitly, we assume that everyone contributes equally – but that is not necessarily true.)

(2) The presentation's clarity. (*Comment:* They must comply with the instruction. Please also pay particular attention to discussions about "lessons learned," if any. Also, look for prototyping and test plan.)

(3) The quality and marketability of the design, or at least its potential. Here, also we need to look for the relevant prototyping and test plan.

(4) The superiority of the selected method – if more than one method is given.

(5) Finally, is the team on a right track to complete the task by one week before the end of the second term. We need the last week for administrative work to finish the course and its final presentation.

Comments to the faculty: Thank you for your assistance, and please feel free to make your observation known in order that we adjust this process as deems necessary.

Final Thought

For many of you, this is perhaps the first experience to stand in front of your peers and professors to talk about anything. However, this is not just anything, it is rather a project that you are working on for several months closely. Thus, by now you should know it well enough to talk a few minutes about it from your heart. Make sure that your presentation conveys that fact. It must also show that you have mastered all elements of this course as outlined in these lecture notes.

Closure

The class ends with the following list of reminders, although many aspects of these issues are included in the above advisory evaluation form.

(1) Practice your presentation in order that your team can complete it in 15 minutes, we will be timing it.

(2) Arrange among yourself an order such that each person talks for a few minutes.

(3) During your presentation, look at your slides and the audience, while explaining your entire project as clearly as you can. As stated in class, you may choose a ***"Big Picture"*** approach, and give a ***"Top Down"*** presentation by defining the overall organization of the project and the role of each member, all the way down to whatever technical details you want to explain.

(4) You must decide, within the allotted time, how much technical detail you must include in your presentation. Note that it is not trivial to say the right amount, and to demonstrate such an understanding and training is a major part of your presentation, which your grade depends on that.

(5) Do not forget to state your ***technical problem*** and ***design specification***. You must convince the audience that your selected method (often there are more than one method or solution for most engineering problems) is the best approach and most likely to be completed on time.

(6) It is important that you give a brief explanation of the qualitative and quantitative aspects of your design in a manner that everyone can understand your way of selecting the solution approach and parts on your project.

(7) Talk about the cost and budget as much as you can at this stage.

(8) Do not neglect to talk about other elements of design as pertains to your project.

Keep these thoughts in mind, as we advance toward the final presentation, and add more elements to this list. *A successful design project is the one that works and is built cost effectively.* Finally, if the slides that you may have dropped earlier in the course instructor's office need major changes, based on the above comments, please bring a revised draft before the start of these presentations. Please be advised that if you decide to change your slides those changes must be consistent with your report. When presentation starts, we will not allow you to replace or adjust your slides based on others' slides. Next week, during our regular meeting, the results of your presentation and the comments on your intermediate final report will be given to you.

Essential thoughts in this lecture

Issues.	Applicability to your project, if any.
Each and every piece of these lecture notes forms a tile on the final mosaic of this course. Try to assemble those such that it portrays your contributions to the underlying project, and your learning of issues discussed in this course. You may wonder, why are we repeating the above? The answer is very simple. It is that serious, and we cannot say it any better than the above!	Obvious! However, you may claim that, for instance, your project has nothing to do with certain issues discussed in class. Clearly, if that is the case, then you must justify your claim.
Do you want to add anything else?	Please elaborate.

To assist you with a potential checklist, which you must develop eventually, we offer the following. This is a partial list of issues that you are expected to answer at some point in this course. Please feel free to add to this list.

> How duties were shared?
> Did they explain their problem statement well?
> Did they look into various solutions?
> How did they pick their solution?
> Did they analyze that solution?
> What is their technical approach?
> Are cost breakdown and budget included in their talk?
> Are schedules and planning with critical dates included?
> What special resources are required, and who would be responsible for procurement?
> Have they included a test plan?
> Did they consider various design constraints?
> Have they evaluated the impacts of other design elements as described in these lecture notes? For instance, Lecture Note Eleven.
> Last, but not least, have they cited sufficient number of references in their talk and final report?

Good Luck!

LECTURE THIRTEEN: Mid - Course Design Review Results

Remark: Clearly, this lecture should reflect the real experience of the course instructor as pertains to a given program. The following however, are what we faced at the University of California at Riverside during the Spring of 2001.

Introduction

The general results of Mid-Course Design Review are discussed in class. It should be pointed out that despite repeated announcement some students have left out important sections in their presentations. The following issues, left out by some, are very important and must be included not only in the final report, but also we believe that in due time in the final presentation.

For Final Report

We shall not repeat the extensive instructions, from the lecture notes on various issues related to this class. However, please be advised that this is a design course and by definition is a culmination of your entire undergraduate program. Every project must have discussions on all issues described in Lecture Notes One, Four, Five, Six, and Seven so far, and whatever else we will be adding to these notes regarding the final report. What some of you have left out in your presentations and must be included in your final reports are as follows. *Note however that, how, and in which sections to describe these items are given in the notes.*

Problem Statement, you must explain in technical terms whatever you are trying to design. Please do not assume that your problem is well known or everyone knows that. Do not assume that others can appreciate the complexity of your project.

Design Specifications, every project has a set of specifications, which are generated based on the problem statement and all relevant design criteria. These specifications in whatever shape and form, numeric or otherwise, must be defined in a manner that leave no ambiguity. You cannot design without a proper set of specifications. Also derive design constraints based on the problem specifications. For instance, if you want to maneuver a 3Kg object that has a specific geometry, then this requirement translates into a specific range of torque that has to be generated by a motor, which in turn translates into a specific operational range for voltage or current corresponding to that motor. Based on this quantitative analysis you select a motor, *per se,* that meets these specifications. This choice, *or lack of it,* inherently imposes a

Design Constraint, which means, one must address or anticipate a set of realistic design constraints when the design specifications are **not** met. These constraints can be on the algorithms or hardware.

Technical Approach, the final report must include this section.

Cost, as imbedded under the section in *Budget,* must be included in the final report.

Test Plan, describe test procedures and results to verify the design and should be included.

Lessons Learned, is the unavoidable part of any design process and must be addressed.

Elements of Design, referring to all issues that are described in this course such as marketing, ethics, which inherently impose design constraints on every engineering project, and these issues must be analyzed and reviewed in the final treport, together with many more that will be described in future lectures.

Again, please note that the above list is only a reminder of issues that were left out by some (in our class), and is a partial list of issues, which you must address in your final report. In other words, to prepare the final report one must include all sections that are described in the notes, while paying a particular attention to the above list.

Faculty Comments on Past Presentations

Here again, the course instructor must provide specific suggestions to students in a given program. In our presentation, one major issue that was raised by the faculty members was that some of the students just read from a piece of paper, which shows they have not yet mastered the subject to the point that they can talk about it from their heart. This situation must be rectified for the final presentations. Also, do not switch back and forth between speakers in your presentations.

Overall, our class did an excellent job, and we had corrected all our problems for final presentations.

Other Lessons Learned

(1) Based on administrative experience, we believe that it is better to request *two* copies of intermediate report, instead of one initially suggested. One copy remains in the student activity file for each group, and the other copy will be returned to students with possible suggestions.
(2) Another issue that must be noted is the following. Some groups changed their slides as they watched earlier presentations. In this process, unfortunately they forgot to change their corresponding report – a situation that must not occur. In future we will be more restrictive about handing over slides.
(3) Despite major efforts to accommodate everyone, a few groups did not benefit from the presence of their advisors. In future, perhaps we must schedule this presentation event as early as the time when projects are selected and the list of participating technical faculty advisor becomes known.
(4) Overwhelmingly, students agreed that the best time for this presentation was during the first week of the second term, because they benefited from the time spent on their preparation during their break.

Additional Presentation Issues

Again, this is our experience at University of California at Riverside during the Spring of 2001. We believe this information may be helpful for the future classes and that is why we are sharing those in this textbook. In our Memo to students regarding their Mid-Course Presentations, we gave specific instructions to them to consult with their technical faculty advisors regarding their preparations. However, some faculty members were indicating that they have not seen the slides and/or the students have left out important information in their presentations. Please meet with your technical faculty advisors weekly, to discuss technical issues and ask them to review your final presentations' slides.

The following list is the essence of the comments, which we had received during our Mid-Course Design Review, and should be carefully considered by all students.

(1) Students should not complain about, nor blame, the hardware given to them. All equipment malfunction should be reported as early as possible. Identify alternative hardware or approaches as early as you can.
(2) Do not be informal, and stress the fact that you will complete the project. You will not be allowed to complain about anything and in particular these peripheral issues during your final presentations.
(3) Do not lose sight that this is a ***technical presentation*** and not a made for TV production.
(4) Do not focus on minor details, and do not assume that everyone knows your problem. The real challenge is to keep a balance of how much or how little you must talk about an issue. Learning this balancing act is both an education and an art!
(5) Again, present your problem statement and specification list clearly and explicitly.
(6) Those of you who discussed cost should know that one-time (non-recurring) cost is different from per- unit (recurring) cost and we will discuss this matter in Lecture Seventeen.
(7) All graphs and figures as well as tables and charts should be labeled properly as discussed in next lecture.
(8) Clearly state who did what in your future presentations and reports.

Final Thought

Although we are open to any last minute ***super performance*** by any one of you, we are quietly compiling a grade based on your performance thus far, which is based partially upon the following factors. The logging in your laboratory notebook, your recent presentation and intermediate reports are among those factors to be considered in this process. These factors will be incorporated with those corresponding to final presentation, and final report, to come up with the final course grade. Certainly, your attendance and other matters discussed in class are also being monitored and incorporated in the final grade.

Closure

The class ends with the following remark. You must build eventually a project that works. So, please do not try to be too ambitious and extend yourself very thinly in several directions. There is nothing worse than saying what went wrong, instead of saying what you have accomplished. Please do not go that way. Or in the words of Professor T.J. Higgins, "Never explain why you didn't do it, no one is interested in that!"

Essential thoughts in this lecture

Issues.	Applicability to your project, if any.
Each and every piece of these lecture notes forms a tile on the final mosaic of this course. Try to assemble those such that it portrays your contributions to the underlying project as best as possible, and your learning of issues discussed.	Obvious! However, you may claim that, for instance, your project has nothing to do with certain issues discussed in class. If that is the case, then you must justify your claim.
Anything else?	Write about it.

LECTURE FOURTEEN: *Technical Writing – Phase II*

Remark: This lecture addresses additional instructions that must be considered to write technical documents.

Introduction

Welcome to the uncharted world of creativity! Certainly, not everyone wants to pursue a graduate study or works as a scientist in a research laboratory. We know all that, but you still need to write about your design and you need to document that, in order to finish this course and many other similar assignments in your career as an engineer. As described in Lecture Note Five, your goal should not be to write such that you are understood. That is not enough here. You must write such that you are *not* misunderstood. To that end, we need to follow a certain structure. One such structure is described below.

But first, note that any technical document displays two distinct characters. One is the actual technical contribution, which must be described in a certain structured fashion. One is suggested subsequently. The other character is the author(s) of the document, which often can benefit from the publication of a strong technical document, although a technically weak document is an equal opportunity destroyer of careers. Thus, be very careful of what you leave behind bearing your name, and do not sell short your good name. Show restraint and discipline in writing, and do not settle for publishing a document or an article, unless you are absolutely certain of its accuracy and authenticity. A strong technical document must be authoritative and substantially add to the state of its discipline. Of course, you need to practice writing, and a few weak papers are acceptable until you master the art of effective and strong (power) writing. In particular, those of you who will pursue a graduate study will write theses and several technical papers, and will learn this process in due time. However, make a habit of being concise and informative from this very beginning.

Although the style and/or procedure to document and structure a technical contribution is different from one field to another, the following sample is what most IEEE Transactions prescribe, and we suggest that you adopt this structure as your *writing template*. We review this sample first, then suggest how to use that to write your intermediate reports for this course.

Please pay special attention to the right or left paragraph, center or left-hand side of the page.

Technical Writing – A Template

(*Comment:* The title must always be direct, short, informative, and without any acronym or any unknown, i.e., you must select similarly.)

Mansour Eslami
Visiting Professor (this is optional)
Dept. of Electrical Engineering
University of California
Riverside, CA 92521
Eslami@ee.ucr.edu

(*Comment:* Name and affiliation of all authors are given here, in the order of their contributions.)

(***Please note well!*** Unless stated clearly otherwise, the implication of the above name is that, this is the author of the following document in its entirety. If there are several authors, as is the case in the Senior Design Course, then the implication of not spelling out specific name in each of the following sections is that each author has contributed equally into this document. However, in the case of multiple authors, when for instance, a particular section is written solely by one person, that person's name must be given explicitly.)

Abstract – In this section a concise account of technical contributions of this document is presented. (***Comment:*** Absolutely nothing else is written in this section. Nothing means nothing – no acronym, no quotation, and nothing about the rest of the document. Just the outcome and goal of your problem statement. Also, note that this section has no number and generally starts at the same line as "Abstract" itself. Some journals put this word in the center and some do not use it at all, but still they require having this section without the word Abstract. Some journals use *Summary*, and some use *Synopsis*. In your final report, we call this section *Executive Summary*.)

I. INTRODUCTION

(***Comment:*** Most IEEE Transactions use Roman numbers to label these headings. Some Transactions center that as well as the subsequent headings, but we use the left-hand side as shown here. Also note that these main headings are all in capital letters, which is not so in sub-headings shown below.)

The ***first paragraph*** in this and any subsequent section starts on the left-hand side, i.e., from the beginning of the line as shown here. The next paragraph is indented as follows.
    Introduction means just that, i.e., in this section you give some clear background information about your technical problem. As far as this course is concerned, you must state your goals for your design, as well as how the rest of the report is prepared, which, in effect, is described below.
    It is a good idea to have this section divided into several sub-sections – identified **A, B, C, ..., *etc.*.** For instance,

A. A Historical Perspective

[Note that we start with "A" historical perspective meaning that we are not covering "Every" historical discussion about the subject – it is safer that way.] Then, we continue by writing that, for instance, since the turn of the century researchers and engineers have worked on this problem [Ref. 64, the style for citation is described in Lecture Note Four, and is placed at the end], but until recently no one knew how to formulate the source of noise in the corresponding circuit, and thus no definitive results were known [Ref. 59].

(***Comment:*** You need to cite references when you speak of other work in the field.)

B. Technical Review

There are many other equally useful methods available to solve this problem such as the recent results published in [Ref. 32] to [Ref. 41], however, our technical approach has many advantages over these methods as described in Section II (referring to your document).
(***Comment:*** Here, you need to make a judicious decision on how much technical review you must present. If your problem has no history, then you have one set of constraints to meet, in order that you convince the

referees that you are doing something worthwhile. On the other hand, if your problem is well known and many people have worked on that, then you really have to show that you have studied those other methods and still are convinced that your approach is better than those other well-known approaches.)

C. Other Pertinent Issues

Depending on the nature of your problem, you may or may not need to have this or any other sub-section here. If you do, then figure out how to label that. It may be a very good idea if you give your problem statement here, but we prefer to keep this section as light as possible, i.e., no equations in this section.

<center>Please note that we are still in Section I (Introduction).</center>

(In the last indented paragraph of this section, starting as shown here, you may write the following, which is sort of a table of contents. In your final report, most of this paragraph is replaced by an actual Table of Contents.) The organization of this document is as follows. In Section II, the technical results are presented. In Section III, the simulation results and the corresponding computational issues are discussed. Experimental results are given in Section IV. (Note that we switch the words around). Section V discusses odds and ends. Section VI to Section before "last" are defined similarly. Conclusions are deferred to Section "last", followed by references, which normally has no section number. If the document has appendices, then the preceding sentence is replaced by the following. Conclusions are deferred to Section "last", followed by several appendices. Appendix A discusses "one issue," while Appendix B discusses "something else," then label the rest of appendices similarly. Finally, you write that, the document ends with the list of references. In your final report, where we use an actual Table of Contents, the above paragraph is not needed in this form, however, you must write something like that in the beginning of each section of that report as a ***"local table of contents,"*** and write that in words.

Please also note that, the entire document must be written in the ***present tense***. Absolutely, no "will be," or "did" is allowed, when referring to a ***location in the body of the document***. When referring to the ***time of occurrence of an event,*** you can say that "we will continue this study next year," or "we did study this problem for several months." But when you are referring to a location in the body of this document, you write, for instance, as described in § I.B, this problem is such and such.

II. TECHNICAL RESULTS (or any heading, depending on the document)

In the first left-hand side paragraph, you give a brief account of the purpose of the document, in perhaps several sub-sections. Followed immediately by the formal problem statement, given in its own sub-section as follows.

A. Problem Statement

This could have been stated informally in the above introduction, but the formal representation must be given here again. Since this is used by senior design students, the above should be followed by a sub-section on ***Design Specification.***

Continue in this section as you deem necessary to explain each subsystem with its schematic diagrams, experimental measurements, or any theoretical argument as you feel is needed. If is justified, then use a new section to describe simulation and experimental results. *You need to know, however, where and when to stop!*

III. Next Section and beyond

Here, and in the subsequent sections describe the rest of your technical contributions. Give a clear description of how various subsections are integrated together and what the limitations are. Keep one issue in mind at every step of the way. ***Be always precise and concise and thus avoid jargon.***

The document continues. Throughout this writing when you need to *label a table,* it is customary to write its description on the top of the table and center that as follows.

Table 2. The results for experiment five.

Date	Monday	Wednesday.	Thursday.	Friday
Results	False	True	True	False

Similarly, when you need *to label a figure or chart,* it is customary to write its description on the bottom and center that as follows.

Imagine this is the frame for an abstract picture.

Fig. 3. Description of a figure.

Note that you may number tables and figures sequentially as above, or within its own section. For instance, if you have three figures in Section III, § **A,** then you may also label them as follows. Fig. IIIA-1, or 2, or 3, which as you see this sort of numbering may soon become cumbersome. Unless, we have used numerical system, such as Fig. 3.4-2, which means the second figure in Section 3, sub-section 4.

On the other hand, when we need *to cite a figure or table,* we use its corresponding number. For example, "as shown (depicted, presented) in Fig. 2, such and such is true." You may also write this as (cf., Fig. 2), or just ", Fig. 2." That is, to put at the end of a sentence, Fig. 2. Never, ever write "see Fig. 2." Not in this course!

"Last number" CONCLUSIONS

You close your document by writing the following. In this document we have shown that all we have set out in our problem statement are achievable and we have completed our task. We also propose the following suggestions for further study or improvement of this issue or project.

(***Comment:*** Here, you confirm that you have finished the job successfully, i.e., what you had set out in the Abstract and/or your Problem Statement. Also, note that it is perfectly acceptable to use "we," throughout this document, even though the author is only one person as herein.)

References (Note that no number is assigned here, and not all letters are capital.)

The references are described following the above section, unless there are appendices, which are described first. The style to cite references has been already shown in Lecture Note Four. You must include some references that discuss similar issues in any technical document, in particular, your final report. This inclusion is imperative for the success of your document.

ACKNOWLEGEMENT

(***Comment:*** List the names of people or organizations that have supported this project in one form or another.)

The template ends here.

Final Thought

Certainly, you must give as many subsections as you need to describe your entire contributions, within the above style, but you must follow the procedures, which we have described in this course. Namely to write about the ***Elements of Design,*** as well as ***Design Constraints*** as pertain to your project.

Closure

The class ends with the following comments that we expect everyone to understand and use the above template to rewrite intermediate reports in that context in order to proceed further.
    A few additional comments, regarding the IEEE standard of writing technical documents.
    Avoid opening a sentence with ***any numeral.*** For instance, "19 dollars is an average cost of this circuit," should be written as: "The average cost of this circuit is 19 dollars."
    It is also customary that equation number be given in parentheses and the references in bracket. Thus, it is understood that (2) means equation number two, and [2] is the reference number two, and there is no need to write eqn. (2), or Eqn. (2), and similarly Ref. [2].
    Finally, the correct way to write numbers, from one to ten is just that, i.e., in English, but from eleven and above is numeric 11, 12, *etc.*.

Essential thoughts in this lecture

Issues.	Applicability to your project, if any.
Each and every piece of these lecture notes forms a tile on the final mosaic of this course. Try to assemble those such that it portrays your contributions to the underlying project as best as possible, and your learning of issues discussed. You really need to think about this template and how to utilize it for your project.	Obvious!
Do you want to add anything else?	Please elaborate.

LECTURE FIFTEEN: In Preparation for the Final Report

Remark: This lecture covers an essential template to prepare the final report of the course. It is presented here, because we are almost three quarters into the course and students must begin putting their thoughts together and organize their exit. The following order is important and must be maintained, but there are flexibilities in various subsections, which can be used as one wishes.

Introduction

The final written report is the crown jewel of this course. This report must cover all the materials presented in the final oral presentation, plus any additional supporting information, which pertain to each project. In other words, everything that is claimed in the final oral presentation must be supported in this document by *technical reasoning*. At this stage, there are two major issues confronting us:

(1) Who does write the final report?
(2) How this report is to be written?

The answer to the first question is obvious. ***Everyone must contribute to this writing.*** All of you must show that you have learned how to write at least a short technical document, per last lecture note. Here is the place to put that training into the use. We suggest that each member writes, say a chapter or two according to the instructions of the preceding lecture note, then try to integrate these pieces by proof reading each other's writing into one final coherent report. You may divide this writing by actually assigning each section to the person whom has actually done the work or has contributed more to that section than others. Each chapter must look like a technical document in the style of the preceding lecture note. The second best thing to do is to let one person write the first draft, then the next person edits and adds to that. Clearly, you cannot write together, in the sense that every line is written simultaneously by every team member. Thus, try to break the work by the sub-area of the final report and let everyone writes a piece, then integrate all pieces. ***Please be specific in identifying the name of the person who has written the specific part of the report.*** Similarly, the person who has led the integration of all parts into the final report should be identified. Finally, you must write this report first, and before preparing from it your slides for final oral presentation, and not the other way around.

No matter who does what, the final written report must be perfected to cover all aspects of the design as described in this course completely.

On the second question of how to prepare this final report, we make the following assumptions first.

(1) We assume that everyone is capable of writing, albeit a short, technical document according to the structure that is described and perfected in Lecture Note Fourteen.
(2) We also assume that everyone is fully cognizant of what we have called repeatedly as the ***Elements of Design*** in this course, which really culminates all our efforts.

Now, the question is how do we put these pieces together? To that end, we offer the following structure, which is an expanded version of a short technical report that is described in the preceding lecture note.

The Structure of Final Report

In addition to each team member's and the technical advisor's copies, **four** copies of the final report must be submitted, each in a clear cover with a plastic slide that holds papers, unless you choose to bind in a manner that takes **less space**. These copies are for the department, library, and the course instructor.

1. The Cover or Title page (This is your page "i," when counting, but we do not put this number herein.)

A ("*A*" means, we have room to modify portions of below as necessary) style for this page is shown below:

Final Report of Senior Design Project
Winter / Spring 2001
Professor Mansour Eslami
(**Comment:** The name of the course instructor)

The University of California at Riverside
Department of Electrical Engineering

(letter size 14)
(Space between any two segments should be chosen accordingly.)

Title of the Project – All Words Begin With an Upper Case
(letter size 12, similarly the rest as well)
(**Comment:** This title can be different from the one you started, if your technical faculty advisor agrees.)

Prepared by: Karl Curtis Doering (The *first author* edits the final report, and perhaps has done more.)
Michael Wilson (If the *second author* has worked equally, put this name on the same line as above.)
(**Comment:** Only here you give names according to their contributions to the final report.)

Technical Faculty Advisor(s): Mr. Bob Kelly, Senior Engineer at Maxim Corporation
(**Comment:** If this person is from outside the department give his/her affiliation.)
(*Also, when there are several advisors, list them in the order of their contributions.*)

This project is supported in part by a contract (grant) from Maxim Corporation.
(**Comment:** Here, indicate the funding agency that supports the project, if any.)

Submitted on June 7, 2001 at 5:00 PM

---

2. The Executive Summary (This is your page "ii," but we do not put that number herein neither.)

The (note that here, you must be exact) style for this page is shown below:

> **Executive Summary** – In just **one** page, present the summary of your work by outlining the problem statement, the chosen method, key features, method of solution, evaluation method, and your important results. You may use several paragraphs, each for one thought, but no more sub-sections. This is just an expanded version of the abstract of your intermediate reports. The last line (or two lines) of this page should look as follows. (Pay special attention to the style, each keyword is separated by a comma or a semicolon; and some begin with upper cases and some do not – here you need to consult your technical faculty advisor).
>
> **Keywords** – *Wireless communications, transmitter design, Wide-band, Code-Division, Multiple Access transceiver design.*

3. Table of Contents *(This is page "iii," which you type at the bottom of this page, and if your table exceeds one page, then the next page is "iv," and so on.) This page is followed by the body of the report as shown subsequently in Appendix A.*

A ("A" means, we have room to modify sub-sections below as necessary) style for this page is shown below:

Table of Contents

Page

Chapter One – Introduction *(This chapter is prepared primarily by the first author, or give the name.)*
(In reality, this chapter is written at the end, when the project is completed and the following is written.)
    1.1. Introduction *1*
*(**Comment:** Before going to the next section, i.e., in the last paragraph of this section, put a local table of contents, describing the rest of this chapter. Similarly, you need to do this for each chapter below as well.)*
    1.2. Whatever Else You Wish To Add *3*
*(**Comment:** For instance, you may give a summary of your problem statement with a historical perspective; motivations and goals of the project; what the general solution(s) may entail; and what the reader should expect to read in this report.)*
    1.3. A Glossary of Acronyms and Abbreviations..*9*
*(**Comment:** Put all such items here in the introduction chapter and refer to this section for future reference.)*

Chapter Two – Design and Technical Results *(This chapter is prepared, for instance, jointly.)*
    2.1. Introduction (follow our earlier instructions for these sub-sections)............12
    2.2. Problem Statement 14
    2.3. Design Specifications 15
    2.4. Whatever Else You Wish To Add (For instance, Lessons Learned)..............16
*(**Comment:** In each of the above sub-sections, you may use block diagrams to break the problem and its various tasks into a few smaller and perhaps more manageable segments, which are delivered by different team members.)*

Chapter Three – Method of Solution *(This chapter is prepared primarily by Michael Wilson.)*
    3.1. Introduction...17
*(**Comment:** Here, you provide an overview of the design solution – the "Big Picture." You may give another sub-section between here and below to review possible solutions, which you did not select. Also, do not forget the Comment below § 1.1. in above.)*

Part Two: Lecture Fifteen – In Preparation for the Final Report   Page 75

Table of Contents Continued

                                                                                           Page

3.2. Our Design……………………………………………………………………*18*

(***Comment:*** *Here, you must emphasize the key features of your special contributions to this design problem. You may use block diagrams to describe each component and its special function. Present whatever you deem necessary in order to give an excellent description of your work. Put your hearts into this section.*)

3.3. Alternative Approach / Design Trade Off ……………………………………...*19*

(***Comment:*** *Presents concept generated and your selection method(s). Those of you who claim are working on an original design and concept should give some explanation here. If details are distracting, you may use portions of those in an appendix.*)

3.4. Lessons Learned ………………………………………………………………...*20*

Chapter Four – Evaluation (*This chapter is prepared primarily by Karl Doering.*)

4.1. Introduction……………………………………………………………………*21*

4.2. Discussion of Results / Test Plan……………………………………………...*22*

(***Comment:*** *This title is just too generic purposely at the moment, however, you need to have a description of your prototype relative to the production model. You need to explain your test plan here, or if you wish so immediately under a new sub-section followed.*)

4.3. Design Comparison / Design Trade Off……………………………………….*24*

(***Comment:*** *In this chapter you need to give a comparison between what you have designed versus your original design specifications. You must state all characteristics of your design, which include its strengths and weaknesses. You may also state your design trade off had you chosen another method of solution.*)

*The above evaluations are relative to your work, in other words, these are **local** and technical evaluations. The **global** evaluations, that compare your entire work with others, are discussed in Chapters Six & Seven.*

Chapter Five – Administrative (*This chapter is prepared jointly.*)

5.1. Introduction……………………………………………………………………*25*

5.2. Budget and/or Cost Analysis…………………………………………………..*26*

(***Comment:*** *Have you met your design specifications as far as the cost is concerned? This evaluation is different from the previous one. There, you can promote your technical contributions by saying our test plan shows that we have met all our technical design specifications. Here, you give a cost analysis, before we go to the next chapter and evaluate the overall project from a number of issues – cost included - globally. This sub-section should include whatever issues are discussed under funding in various sections of this course, more specifically Lecture Notes Eight, Nine, Seventeen to Nineteen, as pertain to your project.*)

5.3. Whatever Else You Wish To Add……………………………………………..*27*

(***Comment:*** *For instance, you may write about your experience in working together, some of your ways of breaking various tasks among yourselves and complying, or otherwise, with all scheduled assignments. Whatever experience you think is relevant and worthy of discussing or sharing with others.*)

Chapter Six – Meeting Expectations (*This chapter is prepared jointly.*)

6.1. Introduction……………………………………………………………………*28*

6.2. Design Constraints……………………………………………………………..*29*

(***Comment:*** *Have you met **all** your design specifications, this is different from Chapter Four. There, you can promote your technical contributions by saying that our test plan shows that we have met our design specifications. Here, you must be explicit in discussing your deviation, if any, and its corresponding effect on the final outcome. Some sort of a sensitivity analysis.*)

Table of Contents Continued

                                                                                                                 Page

      ***6.3. Elements of Design*……………….....………………………………………....30**
(***Comment:*** *All those issues, which are discussed in this course, including, but not limited to, economics factors, safety, reliability, aesthetics, ethics, and social impacts, must be addressed explicitly herein. You may use a table to respond briefly about some of these matters that are tangentially related to your project. However, to neglect addressing these issues specifically will affect your final grades extremely.)*

In the above chapter and in some instances in the next chapter you give a global evaluation of all issues pertaining to your project.

Chapter Seven – Conclusions *(This chapter is also prepared jointly.)*
      ***7.1. Introduction*……………………………………………………………………..31**
(***Comment:*** *In this chapter you explain* **positively** *that you have accomplished your goals to the best of your ability, nothing more, and nothing less. Failure to state in this manner means something is wrong. You are not allowed to complain and give reasons why you have failed. However, if you have not met your goals you must state so, without any request, i.e., do not ask for mercy!)*
      ***7.2. Expansion and Improvement*…...……………………………………………32**
(***Comment:*** *Discuss the impact of this work and its possible expansion into perhaps a more promising design than what you had started. This is particularly important in order to address the marketability of your design. Or why this project merits another look by perhaps next year's students.)*
      ***7.3. User's Manual*…………………………………………………………………..33**
(***Comment:*** *If your design requires instructions for future use, here is the place to put that information.)*

***Appendix A: Parts List*……………………………………………………………….34**
***Appendix B: Equipment List*………………………………………………………...35**
(***Comment:*** *Some of the special hardware components and/or equipment, which are used in your design must be described in more detail than what is presented in your intermediate reports. Choose this place to explain those issues.)*
***Appendix C: Software List*……………………………………………………………36**
***Appendix D: Special Resources*…..…………………………………………………..37**
***Appendix E to Z: Whatever Else You Wish To Add*………………………………..38**
(***Comment:*** *For instance, here, you may include detailed solution methods or derivations, which you need for your future review of this report, or whoever else is interested to pursue this study. Some side drawings and printouts that are of value to people who may continue this work, should be given herein. Some information about the vendors and how to locate parts for similar projects must be included herein. In other words, information that is important about the overall construction of the project should be given herein.)*

***References*………………………………………………………………………………39**
(***Comment:*** *Use our citation procedures and list in alphabetic order all your references based on their first, second, etc., authors, in a chronological order. You may include these references depending on each chapter or as a whole. A report without references is considered very weak and unacceptable. A report without references from a refereed media is considered weak.)*

***Acknowledgement*………………………………………………………………………40**
(***Comment:*** *Be considerate and credit all those who have helped you in this project, especially, if your advisor have paid for the expenses from his/her grants.)*

Final Thought

Although we have several more weeks of school, we put these instructions before you, in order that you start working on your final report. We hope that you use the above template together with your own unique style to come up with a very successful report. We also hope that you enjoy using the above for all your future technical writings.

Closure

The class ends with the following comment that each person in any given team should look into one piece or mosaic of the above "Big Picture," and get every thing ready before the deadline.

Essential thoughts in this lecture

Issues.	Applicability to your project, if any.
Compartmentalization of thoughts finally pays off!	Obvious!
Do you want to add anything else?	Please elaborate.

Appendix A: The Checklist for Final Report

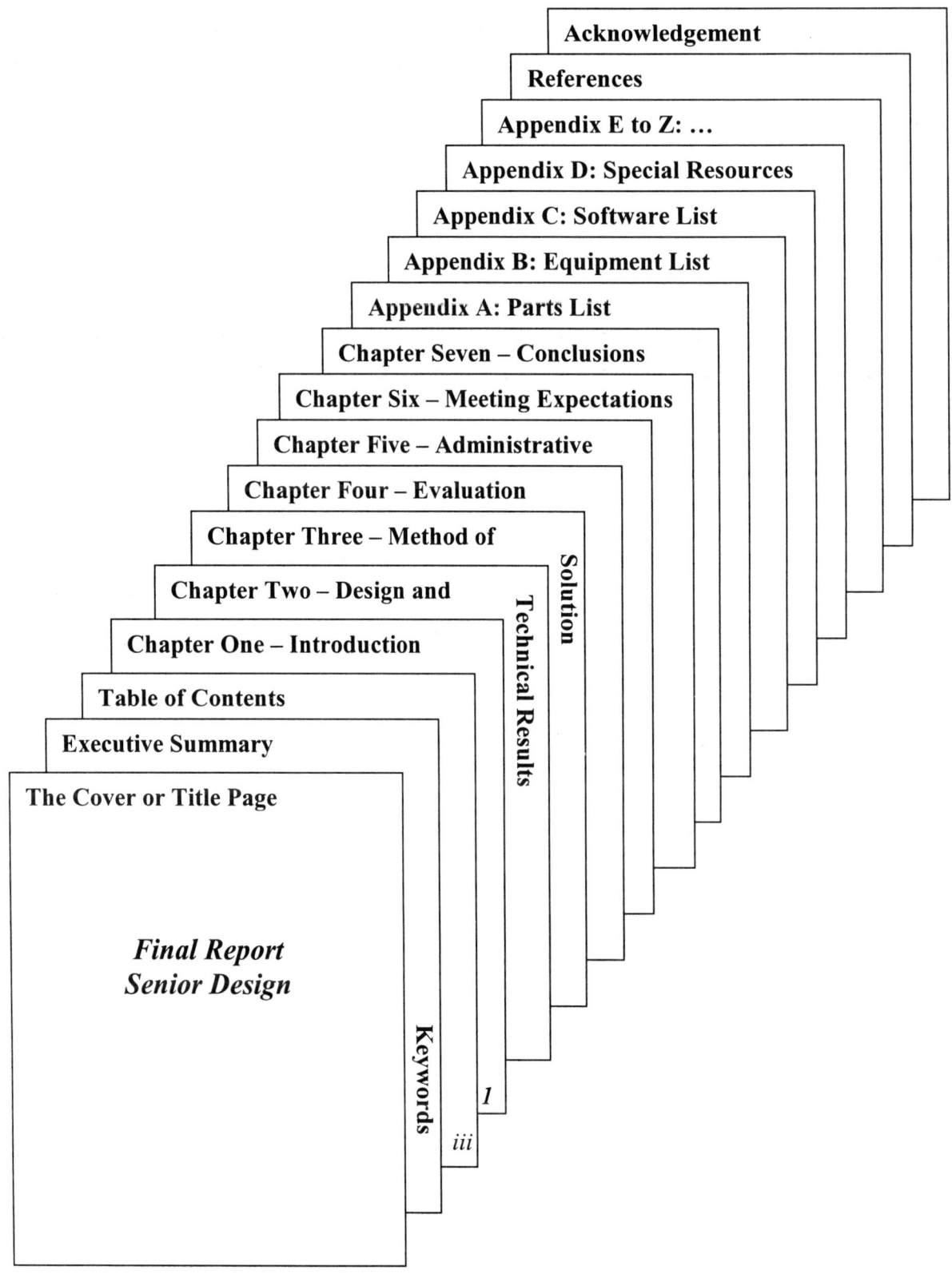

LECTURE SIXTEEN: *Steps Towards Oral Presentation – Phase II*

Remark: The oral presentations are of major concern to us, and thus we are taking time to schedule another class-wide presentation, but this time only for students. The faculty members are not invited formally, although they can participate if they wish so. This event can be scheduled at the convenience of all parties involved, and therefore it could be given here or be replaced within the next three lectures.

Introduction

Students should be briefed on the procedures of their next oral presentations, which the time and date are selected. This presentation is very similar to the last one, i.e., we still allow only five slides, and 15 minutes per group, to demonstrate the technical aspects of their design to date. Also, they must show that they are on the right track and are pursuing a genuine design project. We expect that they have upgraded their materials in order to reflect the current status of their project as well as to incorporate all the suggestions we have made following their last presentations. In other words, to repeat the same mistakes here will be totally unacceptable. These presentations should be video taped, if possible, for future students, and review of the current ones, in order that they can improve their presentational skills and get ready for their various oral presentations.

Students have also been informed that they will be asked to rank each other's presentation, based on a questionnaire that will be distributed on the day of presentation and its essential elements are described next. Again, we must emphasize that this is only an advisory evaluation. But more importantly, this is the same evaluation form (with last question removed) that the faculty members, who will attend the final oral presentations, will use to rank each project. Therefore, you must concentrate on these questions as the initial set of questions that you are responsible for in your final defense.

Design Review Questionnaire

Please rank each item below on a scale between *0* to *10*. This evaluation (based on the lecture notes in this course) will be used only for helping your fellow students to get an idea about their performances so far. We reserve the right to adjust accordingly, and based on many other factors, this recommendation if we will incorporate that in the final course grade. Please use *N/A* when you think that an item is not applicable to the given presentation (that is different from not stating, which ranks *0*). Please also note that we still have quite a few more lectures to deliver before the end of the course, therefore consider that fact.

1. How well did the group describe their basic *technical* problem?
2. How well did the group describe their *technical approach* of solving this problem?
3. How well did the group describe the *organization* of tasks involved and their *shared duties*?
4. Did their *block diagrams, schematics, or photographs* shown demonstrate their design adequately?
5. Rank the quality of their *analytical work* shown relevant to the system or any of its subsystems.
6. Rank the quality of *actual data* taken in the laboratory.
7. How confident are you regarding the quality of their *researched alternative solutions* as presented?
8. Did they *thoroughly research* their problem with all its possible solutions?
9. In other words, if a group claims that this is an *original work,* then how convincing do their arguments

sound to you? For instance, what are the potentials for the **patentability** and/or any **intellectual property** of this work?
10. How confident are you regarding the **theoretical or analytical results** presented?
11. How confident are you regarding the **simulation results** presented?
12. How confident are you regarding their **test plan**?
13. How confident are you that the **final design will actually work**?
14. How confident are you that the final design will be **cost effective**?
15. How well have they met their **design constraints**?
16. How well have they described the pertinent **elements of design** in this project?
17. How well did they **utilize lessons learned** in this project?
18. How well have they promoted their design for a **future study**?
19. In other words, what are the **marketability** and **patentability** potentials of this project? Is there anyone interested to invest in this project?
20. Rank the **overall technical quality** of their entire presentation.
21. How confident are you that this group will have **a working and complete project** by our deadline?

Final Thought

We are getting close to the end of this course. Please make sure that your project is in **working condition,** and you are about to **complete all documentation** for the final presentation and report.

Closure

The class ends with the following remarks. We have the capacity to forget and forgive past mistakes of all of you, but make sure that your final oral presentation and report are perfected, if you are expecting a very good grade. We then return the evaluation forms to each group and ask them to review the comments made by their peers and tabulate their scores and bring the results to the course instructor for a final tally. The course instructor can then announced the overall performance of the class in future lectures as time permits. We have found students tremendously benefiting from this exercise and we recommend highly that approach for the future classes.

Essential thoughts in this lecture

Issues.	Applicability to your project, if any.
Make a list of your observations during this mid-course design review for your own final presentation.	Obvious!
Do you want to add anything else?	Please elaborate.

Good Luck!

LECTURE SEVENTEEN: Engineering Economics

Remark: In this lecture aspects of engineering economics are discussed. First, a formal set of definitions for various forms of "cost" is given, followed by a business plan/cost analysis. Secondly, two standard problems of calculating compound interests for loans and investments are described.

Reference:

[1] S. Davidson, M.W. Maher, C.P. Stickney, and R.L. Weil, *Managerial Accounting, an introduction to concepts, methods, and uses.* Chicago: The Dryden Press, 3rd edition, 1988.

Introduction

In preceding lecture notes we have used repeatedly the word "cost" in one form or another. For instance, in Lecture Note Twelve, we have stated that "A successful design project is the one that works and is built cost effectively." We have not, however, defined the word "cost" as financial experts do that. According to [1], the term *"cost"* is defined as "The *sacrifice*, measured by the price paid or required to be paid, to acquire goods or services. The term "cost" is often used when referring to the valuation of a good or service acquired. When "cost" is used in this sense, a cost is an *asset*. When benefits of acquisition (the goods or services acquired) expire, the cost becomes an *expense* or *loss*. Some writers, however, use cost and expense as synonyms. Contrast with *expense*." According to the above reference, some of the definitions used in this course should be considered formally as follows.

Cost sheet. "Statement that shows all the elements comprising the total cost of an item."

Cost estimation. "The process of measuring the functional relation between changes in activity levels and changes in cost."

Cost allocation. "Assigning costs to individual products or time periods. Contrast with cost accumulation, referring to bringing together, usually in one single *account,* all costs of a specified activity."

Cost / benefit criterion. "Some measure of *costs* compared to some measure of *benefits* for a proposed undertaking. If the costs exceed the benefits, then the undertaking is judged not worthwhile. This criterion will not give good decisions unless all costs and benefits flowing from the undertaking are estimated."

Cost of goods manufactured. "The sum of all costs allocated to products completed during a period; includes *materials, labor,* and *overhead.*" [Please note that both labor (your time) and overhead (the university facilities, *per se*) are included in this definition of cost.]

Cost effective. "Among alternatives, the one whose benefit, or pay off, per unit of cost is highest. Sometime said of action whose expected benefits exceed expected costs whether or not there are alternatives with larger benefit/cost ratio."

Cost behavior. "The functional relation between changes in activity and changes in cost. For example, *fixed* versus *variable* cost; *linear* versus *curvilinear* [not linear, but a continuous curve] cost."

From our point of view, we must look into a *non-recurring* or fixed (one-time) cost versus *recurring* or variable (per-unit) cost. The production cost is per-unit cost when many units are built, and that cost is less than per-unit cost of a prototype model. Here, a host of other costs (storage, handling, *etc.*), however, must be added to the cost sheet, when we produce a large amount in order to reduce its production cost.

In your final report, include a cost behavior (analysis) for your project, paying special attentions to the above items. You may consider your own time as one-time cost of production, thanks to the word "sacrifice," in the definition of cost. Use your logging book to calculate that. We do appreciate your sacrifices!

A Typical Cost Analysis / Business Plan

Prototype cost (These are generic names, you may break these items into more sections.)
- ***Parts*** — $64.00
- ***Software*** — $220.00
- ***Direct labor*** (40 hours @ $25.00/hour) — $1,000.00

(***Comment:*** *This cost refers only to the actual time spent to build and test the first unit. Your salary is not included here, that is considered in the overhead expenses and anticipated profits.*)

- ***Company Facility / Capital Equipment*** (purchase price $12,350.00)
  Depreciation schedule: 40% the first; 30% the second; 30% the third year. — $4,940.00

(***Comment:*** *This is the net expense of using some capital equipment, say for six months, to develop a product. The above accelerated schedule is acceptable by the IRS, although the above ratios change each tax year. The implication of this calculation is that you could have either rented this equipment for one year, or you could have borrowed the money to purchase it, then resell the unit. Just be aware that this is a non-recurring cost.*)

Marketing Analysis / Consultant Fee
- ***Expenses to Conduct a Marketing Analysis*** — $12,000.00

(***Comment:*** *The above study recommends the production of 10,000 units of this item.*)

Estimated Costs to Produce 10,000 Units
- ***Parts*** *(purchased at bulk rate)* — $28,000.00
- ***Labor*** *(10 units/hour/@ $25.00/hour)* — $25,000.00
- ***Storage*** *($1.00/square ft/month) – 350 square ft, for six months* — $2,100.00

Advertising cost — $10,000.00

Total — $83,324.00

Overhead Cost (at a modest rate of 85% for the above) — $70,825.00

(***Comment:*** *To cover capital borrowing, staff salary, insurance, legal fees, office rent, utilities, and all the other costs of running a business. You must even consider the price of the pencil and paper, which you have used to draw the first sketch of this design!*)

Expected Total Profit — $75,000.00

(***Comment:*** *Based on the time spent, and all the other market analysis recommendations.*)

Grand Total — $229,149.00

Expected Sales — **7,800 Units**

(***Comment:*** *When all said and done, everyone knows that not every unit will be sold, also there are unpaid bills, as well as defective units. The above number reflects all these facts.*)

Price per unit, rounded up — $29.99
Shipping and Handling — $6.99

Good Luck!

Compound Interest

One of the inevitable facts of life is that at some point we either borrow money to purchase a "big item," or save money to accumulate wealth for our future expenses and/or purchases. In all these cases, the interest which we either pay or receive are compounded, i.e., interests are added to interests. We like this compounding when we receive it, but not so when we pay it. Banks calculate interest on most installment loans on a monthly, instead of a daily basis. While the actual interest on savings accounts are compounded on a daily basis in many banks. In the following we review two scenarios when we apply for a loan to expand a business, and when we open a saving account to reach a desired level of capital accumulation within a given time, to expand our business. In either of these cases, we rely on our own way of looking at these issues, and derive formulas that we can use in similar cases without relying on any special calculator or various tables and charts. These issues are elucidated in the following two examples.

Example 1: Suppose we want to borrow an initial amount of $X at the rate of β% per year, and pay back this loan in **n** years. What is the *best* strategy to borrow this loan? Clearly, by *best* we mean the least amount of interest paid to the lender.

Solution: There are a few issues that must be clarified before proceeding further. First of all, there are many additional expenses associated with a loan, which are not described in the above. For instance, banks charge an advance interest, which they call "points," and that depends on the size of the loan. For example, if we borrow $100,000 on a 3-point term, the bank may charge us $3,000, in advance, for points. (That means the bank actually pays us $97,000, and calls that a $100,000 loan!) Then there are a host of other fees, which in effect are added to the total cost of the loan. Not only all of that, but also there may be another fee if the loan is paid off early! Say, you decide to pay back a 30-year loan after 12 years, then the bank *may* charge you (some do not) a fee for early payment! So, ask all these questions when applying for a loan. What banks want, and they often get, are a structured payment procedure, which we pay back every drop of interests that they have asked us.

Now, going back to develop a solution for the above Example, let our initial loan (present value) at an annual rate of β% be shown as

$$x_o = \$X.$$

The value of this loan at the end of the first *period* is

$$x_1 = \$X + \$\text{Interest}.$$

How do we calculate this interest depends on the definition of *a period*.

If our loan is set up such that we pay it back in its entirety at the end of its term (final value), say **n** years, then the interest due at the end of each year is calculated as follows.

At the end of first year
$$x_1^{year} = X + \frac{\beta}{100}X = \left(1 + \frac{\beta}{100}\right)X.$$

Similarly,
$$x_2^{years} = \left(1 + \frac{\beta}{100}\right)x_1^{year} = \left(1 + \frac{\beta}{100}\right)^2 X.$$

$$\vdots$$

$$x_n^{years} = \left(1 + \frac{\beta}{100}\right)^n X.$$

This is called a *lump sum* payment, which is not a common practice among lenders, nor most people have the discipline to set aside sufficient funds on a regular basis in order to come up with the above at the end of the **n**th year.

The most commonly used period, practiced by most lenders, is one *month*. Here the interest at the end of the first month is

$$\frac{\beta}{1200}X,$$

where, β% per year is divided by 12 months. Thus the value of the loan at the end of the first month is

$$x_1^{month} = X + \frac{\beta}{1200}X = \left(1 + \frac{\beta}{1200}\right)X.$$

This is how the most loans through banks are set up.

On the other hand, if the above interest is calculated on a *daily basis*, which banks are not supposed to do *for loans*, then the value of the loan at the end of the first day is

$$x_1^{day} = X + \frac{\beta}{36500}X = \left(1 + \frac{\beta}{36500}\right)X,$$

and at the end of the month becomes

$$x_{30}^{days} = \left(1 + \frac{\beta}{36500}\right)^{30} X.$$

Here, we can easily show that for an average month of 365/12 days

$$\left(1 + \frac{\beta}{36500}\right)^{365/12} > 1 + \frac{\beta}{1200}, \quad \text{for all } \beta > 0.$$

To prove the above, we offer the following argument. Let $\gamma = \frac{\beta}{36500}$, then we must show that

$$(1+\gamma)^{365/12} - 1 - \frac{365}{12}\gamma > 0, \text{ for all } \gamma > 0.$$

Expanding the above yields the following, where all terms in + ... , are positive.

$$1 + (365/12)\gamma + \cdots - 1 - \frac{365}{12}\gamma > 0.$$

Thus, when negotiating for a loan, we must plan to pay back on a monthly basis and bank should also calculate our monthly installment accordingly. However, when we collect interest, we must insist on accumulating that on a daily basis, in order that our final *yield* becomes more than the actual annual percentage or rate. The banks, however, know these differences quite well!

Now, going back to the analysis, we choose a monthly installment approach to pay back our loan. The following chart displays the propagation of interests accumulated versus payments rendered throughout the course of this loan, which we call an activity chart. Let our equal monthly payment be called $y. Then

Month	Payment	Value of loan at the *end* of the corresponding month
0	0	$X (annual rate of β% over **n** years, or $\alpha = 12n$ months)
1	$y	$\underbrace{\underbrace{\left(1+\frac{\beta}{1200}\right)X}_{\text{value}} - \underbrace{y}_{\text{payment}}}_{\text{net value}}$
2	$y	$\left(1+\frac{\beta}{1200}\right)\left[\left(1+\frac{\beta}{1200}\right)X - y\right] - y = \left(1+\frac{\beta}{1200}\right)^2 X - \left[1+\left(1+\frac{\beta}{1200}\right)\right]y$
$\vdots$	$\vdots$	$\vdots$
$\alpha = 12n$	Last payment $y	$\left(1+\frac{\beta}{1200}\right)^\alpha X - \left[1+\left(1+\frac{\beta}{1200}\right)+\cdots+\left(1+\frac{\beta}{1200}\right)^{\alpha-1}\right]y = 0.$

The above equation yields **y**, our equal monthly payment. Thus,

$$y = \frac{\left(1+\frac{\beta}{1200}\right)^\alpha X}{\left[1+\left(1+\frac{\beta}{1200}\right)+\cdots+\left(1+\frac{\beta}{1200}\right)^{\alpha-1}\right]}.$$

Using $(a^m - 1)/(a-1) = a^{m-1} + a^{m-2} + \cdots + 1$, we have

$$y = \left[\frac{\left(1 + \frac{\beta}{1200}\right) - 1}{\left(1 + \frac{\beta}{1200}\right)^\alpha - 1}\right]\left(1 + \frac{\beta}{1200}\right)^\alpha X,$$

or simply

$$\boxed{y = \frac{\left(\frac{\beta}{1200}\right)X}{1 - \dfrac{1}{\left(1 + \dfrac{\beta}{1200}\right)^\alpha}}.}$$

Please note that by putting $\frac{\beta}{1200}$ in a parenthesis, we are emphasizing that that quantity should be calculated first, in order to utilize the above formula. We need no chart, no table and nothing else but a simple calculator that computes a^b.

For instance, if we want to calculate our monthly payment for a $100,000 loan at $\beta = 7\%$ annual rate over 30 years, then $\alpha = 360$ months and we need to compute y in the following order

$$\frac{\beta}{1200} = \frac{7}{1200}$$

$$1 + \frac{\beta}{1200} = 1 + \frac{7}{1200}$$

$$\frac{1}{\left(1 + \frac{7}{1200}\right)^{360}} = 0.123205855$$

$$y = \frac{\frac{7}{1200}}{1 - 0.123205855} \times \$100{,}000 = \$665.30.$$

It is to be noted that these early payments cover mostly the interests and very little of the principal.

The above formula can also be used in a reverse analysis. For instance, if we can afford to pay $2000.00 per month, then what kind of loan will we qualify for at, say 8% annually over a 20-year term? The answer is

$$2000 = \frac{\left(\frac{8}{1200}\right)}{1 - \dfrac{1}{\left(1 + \dfrac{8}{1200}\right)^{240}}} X \quad \Rightarrow \quad X = \$239{,}108.58.$$

Part Two: Lecture Seventeen – Engineering Economics *Page 87*

Of course, the preceding formula is a nonlinear equation and in some cases we need more calculations than the above, just as any such nonlinear problem entails. For instance, if we were to ask in the above, find the best rate (β) or the best term (α) such that we can afford to borrow, say, \$300,000 with a monthly payment of \$2000.00. Then we must answer this question by a trial and error approach, using a few different β's and α's. In most cases, however, we have very little to say about the rate, because banks set that up on a daily basis and without any input from us. We need to shop around for the best rate, but by and large that is not in our control. So we start with, say, $\alpha = 360$ months, and a series of different β's. Then we go to $\alpha = 240, 180, 120$ months to match the best loan strategy, which we can afford. Finally, note that our actual total payment is αy. For instance, in the preceding Example, we have paid 360 times \$665.30, which is \$239,508. Thus, in the life of that loan, we have paid \$139,508 in interests!

The following example is another point of view in this overall study.

Example 2: Suppose we want to accumulate \$X after **n** years, by setting aside an equal monthly annuities of \$y. How do we analyze this situation?

Solution: Clearly, the interest rate changes from day to day, let alone from month to month. But for the sake of argument we assume that our rate of return, shown by $\beta\%$ is constant throughout this process. Otherwise we negotiate with the bank. Then an activity chart for this analysis becomes as follows.

Month	Payment	The future value of asset
0	\$y	$X = \left(1 + \dfrac{\beta}{1200}\right)^0 y$
1	\$y	$X = \left(1 + \dfrac{\beta}{1200}\right)^1 y + \left(1 + \dfrac{\beta}{1200}\right)^0 y$
2	\$y	$X = \left(1 + \dfrac{\beta}{1200}\right)^2 y + \left(1 + \dfrac{\beta}{1200}\right)^1 y + \left(1 + \dfrac{\beta}{1200}\right)^0 y$
$\vdots$	$\vdots$	$\vdots$
$\alpha = 12n$	Last payment \$y	$X = \left[\left(1 + \dfrac{\beta}{1200}\right)^\alpha + \left(1 + \dfrac{\beta}{1200}\right)^{\alpha-1} + \cdots + \left(1 + \dfrac{\beta}{1200}\right) + 1\right] y$

Thus, the total asset accumulated at the end of the term in this investment strategy becomes

$$X = \frac{\left(1 + \dfrac{\beta}{1200}\right)^{\alpha+1} - 1}{\dfrac{\beta}{1200}} \, y.$$

To give a numerical example, consider the case where we want to set aside for instance, y = \$2,100 per month at a rate of 5% return, in order to accumulate **X** = $100,000. Then how long do we have to save?

The answer is as follows.

$$100{,}000 = \frac{\left(1 + \frac{5}{1200}\right)^{\alpha+1} - 1}{\frac{5}{1200}} \times 2{,}100,$$

$\alpha = 42.53$ months.

Final Thought

In the preceding analysis β may, of course, change and in that case we must compute the corresponding return accordingly by using several different β's. The real message, however, is clear: We can compute both loans and investment strategies using our simple calculator. Furthermore, if we have the discipline, we can accumulate wealth quickly. Finally, we leave you with the following thought. How would you define wealth? Most people say a rich person is the one who can do whatever he/she wants to do. We define a rich person as someone ***who does not do whatever he/she does not want to do! Are you rich?***

Closure

The class ended with the following remark by Mr. Richard Bleier, who is the author's colleague and has said "everything is valuable as long as you have enough of it." So take nothing for granted and cherish whatever you have. The preceding analyses are helpful tools, which you can use for the rest of your life. Hopefully, when you buy your first home.

Essential thoughts in this lecture

Issues	Applicability to your project, if any
Two important formulas used in computations related to obtaining loans and setting up investment strategies.	Obvious!
Anything else?	Write about it.

LECTURE EIGHTEEN: *Entrepreneurial and Venture Capital*

Remark: The purpose of this lecture is to introduce engineering students to entrepreneurial and new venture issues as well as financing resources. In this lecture, we discuss how to write a business plan and seek a successful venture capitalist (VC) to invest in a startup company.

The following lecture was prepared and delivered at the University of California at Riverside by Mr. Kayvan Mozaffari – a former student of this instructor. The materials covered herein are closely related to materials covered in Lecture Notes Eight and Nine, also delivered by the same speaker.

References:
[1] Publicly available sources such as local business journals.
[2] The Wall Street Journal (http://www.WSJ.com).
[3] The Mercury News (http://www0.mercurycenter.com).
[4] Small Business Administration (http://www.sba.gov), in particular (http://www.sba.gov/starting/indexbusplans.html).
[5] Business Plan Software and Sample Business Plan by Palo Alto Software, Inc., (http://www.bplans.com), and (http://www.paloalto.com).

Introduction

Here, we discuss issues related to high-tech startup companies (same as high-tech ventures), entrepreneurship, business plans, venture capitalists (VC's), as well as other financing resources available to entrepreneurs. A generic outline of a business plan is shown in Section III. There are many different outlines in writing a business plan. These variations are due to differences between businesses. Generic business plans can be found in [4] (specifically in http://www.sba.gov/starting/indexbusplans.html), and [5]. Also, it is to be noted that a business plan is different from a marketing plan, as discussed in Lecture Note Eight. A business plan is more comprehensive than a marketing plan, and includes a short version of a marketing plan.

I. High-Tech Ventures and Entrepreneurial

High-tech ventures usually begin by motivated technical people who have new ideas about a specific technology and product or service. These entrepreneurs have a wealth of specialized information about the current state of technology, suppliers, and customers in their own industry. They also have identified underserved markets, which they wish to serve. To be successful however, high-tech entrepreneurs must be persistent, aggressive, and comfortable with the uncertainty associated with high risk and possible high reward circumstances. They must also keep their technical knowledge current and stay focused on delivering added value to their target customers, which is well differentiated from their competitors.

Entrepreneurs share the following common traits: Entrepreneurs are doers, and can be characterized as people who act swiftly, are decisive, extremely aggressive, and like practical solutions that are easy to implement. The venture capitalists prefer to fund entrepreneurs who have a great deal of devotion to their purpose.

Entrepreneurs – Common Traits

- Act quickly, decisively, and prefer to "just do it".
- Are high achievers and aggressive.
- Are comfortable with uncertainty.
- Are interested in high risk/reward assignments.
- Have high expectations for themselves; always push themselves and others to achieve more.
- Are independent, and want autonomy and freedom.
- Devoted to their purpose.
- Tough minded.
- Like to implement simple and practical solutions.
- Know their values.
- Optimists – the glass is half full not half empty.

Based on the above traits, do you consider yourself an entrepreneur? If so, do you have a slick new idea for a product or service that is the "next best thing"? If so, are you ready; do you have money, partners, and customers to start a new business? If not, then you will need to prepare a business plan to begin communicating your ideas with investors, then to hire key employees and get assistance from suppliers and potential customers.

II. Why Do You Need a Business plan?

A business plan is a tool to communicate and organize your thoughts. It is a starting point for a living document, which shows you think rationally and strategically. It is also a great description of your implementation plan. A good business plan includes a clear roadmap for the company, detailed examination of critical steps, and a comprehensive analysis of the competitive landscape. It must outline the financial and operational milestones, with supporting details for budgetary requirements. It is your brochure to sell your ideas and your company to investors, future employees, and potential stakeholders.

Why Do You Need a Business Plan?

- It is a great communication tool.
- It is primarily used to organize your thoughts.
- Forces you to think rationally and strategically.
- It is a living document.
- It describes your implementation plan.
- Shows a clear roadmap for the company.
- Includes a pros and cons analysis.
- Shows the critical steps.
- Shows the competitive landscape.
- Outlines the financial and operational milestones.
- Details budgetary requirements.
- It is a brochure to sell your company.
- Utilized as a sales tool to raise money.
- Utilized as a sales tool to recruit executives.

III. Writing a Business Plan

The following is a generic outline of a business plan, which is explained in the forthcoming sections.

A Generic Business Plan Outline
A. Executive Summary.
B. Key People and Management.
C. Description of the Problem.
D. Description of the Solution.
E. Intellectual Property.
F. Market Analysis.
G. Sales and Distribution Channels.
H. Financial Analysis.
I. Exit Strategy.
J. Appendix

The starting point in writing a business plan is research. You will need to speak with your potential customers and competitors to evaluate the viability and salability of your idea for the deliverable product or service. You must be honest and gutsy to make the right choices. Determine whether your idea is *marginal* (i.e. not enough advantage for the target customers to buy from you), and if so, can you improve it? Should you allocate your resources (time and money) as well as others to this idea? Or, should you change the model completely and start something new?

Where Do I Begin?
- Begin only if you are serious about your idea.
- Do a great deal of research.
- Speak with your target customers.
- Will your target market buy from you?
- Speak with your competitors.
- Why did your competitors not offer this solution?
- By the time you finish the first draft of the business plan; you must:
  - Conclude its viability.
  - Determine if it is a marginal idea? If so, should you kill it?
  - Decide whether to allocate your time and resources to this idea?
  - Determine if you can save it by changing your model or approach?
- Be honest, be extremely critical, be a realist.

After conducting the research, begin writing your business plan, but only if you are still convinced you should continue. Below are the main sections of a business plan, which are discussed in detail.

A. Executive Summary

This is your best chance to impress the reader. Begin with an overview of the problem and your solution for this problem (do not include any proprietary information in this section). Use well-known and credible information sources to back up your claim. Include how much money is required to implement your plan successfully. Amongst the high-tech market research firms that are well respected in the business community and ready to provide you with market research, we name International Data Corporations (http://www.idc.com), Gartner Group (http://www.gartner.com), and Dataquest (a subsidiary of Gartner Group). However, be aware that no one works with complete information about all facets of a job. Your good judgement must always guide you when you have enough information to tackle a project.

Business Plan - Executive Summary Section

- Executive Summary:
    - One page long.
    - Your BEST chance to impress.
    - QUANTIFY.
    - Information sources are VERY important.
        - Use credible market research companies such as:
            - Gartner Group and Dataquest.
            - International Data Corporations.
    - How much money are you asking for:
        - Do not under or overestimate.

B. Key People

Include a brief biography for key managing directors and founders. If the company has grown already or has been around for sometime, include information about the Chief Executive Officer (CEO), Chief Financial Officer (CFO), Chief Technology Officer (CTO), Chief Marketing Officer (CMO), Chief Operation Officer (COO), and Vice Presidents (VP's) of technical and non-technical people. Also, in the appendix compile (but do not distribute) brief background information regarding board members, scientific and technical advisory board members, business advisory board members, and early investors, if any. Include company name, type, location, and contact information.

Business Plan – Key Employees Section

- People, people, people:
    - Key employees (founder(s), CEO, CFO...)
    - Advisory boards:
        - Scientific/technical board members.
        - Board of Directors.
    - Previous investors:
        - Angels.
        - VC's.

C. The Problem

In this section, identify the problem and reasons for its existence. Focus on a detailed description of the real problem that you have identified. Also, describe why the existing solutions are not adequate, and include an in-depth cost analysis of current solutions. Finally, verify that your target customers agree with your assessments. To achieve this goal, speak directly with your target customers.

Business Plan – Problem Section

- The Problem
    - Where is the pain?
    - What is the current solution?
    - What is the cost of current solution?
    - Does the customer agree with you?

D. The Solution

This section covers your business proposition. Write a detailed description of your solution and its cost. Describe whether your solution is an improvement over currently available solutions, or it is a completely new approach. Show a detailed analysis of your solution with respect to existing solutions, using graphs and tables to rank them. Show all the credible sources that validate your analysis. The main conclusion revolves around superiority of your solution, backed by early indications that your target customers will actually purchase your solution. Briefly discuss the milestones, critical path, and resources required to implement your solution. The details of implementations are covered in Section V.

Business Plan – Solution Section

- The Solution:
    - Your business and value proposition.
    - Its cost.
    - Is it an improvement over existing solution or is it a revolutionary idea?
    - Why do you think the customer(s) will pay for it?
    - Have you credibly validated your solution?
    - Timeline and resources as well as critical path and milestones.

E. Intellectual Property

In this section, describe your possible advantage over the competition. List all your patents, trade secrets, and know how. Include any exclusive agreements that you have with suppliers or distribution channels. Indicate if these agreements create a barrier for competitors to compete with you directly. At this stage you should be working with an attorney to make sure you are making valid claims about intellectual properties, which you legally own. Also refer to Lecture Note Twenty One for more detail on this topic.

Business Plan – Intellectual Property Section

- Intellectual property:
    o What is your possible advantage?
    o Patents.
    o Trade secrets.
    o Know how.
    o Exclusive agreements and contracts.

F. Market, Trends, Competition

A detailed marketing analysis is covered in this section. The guideline for developing such a marketing plan is described in Lecture Note Eight.

In this section, you take a high-level view of the market that you are about to enter. Evaluate marketing issues such as the overall market size and trends, whether your target market is fragmented or concentrated. Identify target markets you intend to penetrate first. Discuss government regulations, the existing value chain, and key competitors, for each target market you plan to attack. In the event you are the first to enter a new target market, how long can you maintain your first mover advantage over the current and future competitors?

In addition, you must cover detailed information about your target customers using the technology adoption life cycle categorization, as discussed in Lecture Note Nine. Given the target customers, you need to describe your unique value proposition and differentiation.

Business Plan – Marketing Section

- Market trends and competition:
    o Who are your customers and why?
    o What is the market size, growth projections, and trends?
    o Is your target market concentrated or fragmented (show percentages)?
    o What is your positioning?
    o What is the current value chain?
    o Who are the key players?
    o What is you unique value proposition?
    o What is your differentiation?
    o Do you have the first mover advantage? If so, for how long?
    o Where are your customers on the Technology Adoption Life Cycle?
    o Who makes the buying decisions?
    o Who influences the buyer?
    o Government regulatory bodies:
        - FCC, FDA, etc…
    o How they interact with you and your market?
    o Do you need their approval?

G. Sales and Distribution

Discuss issues regarding sales price and distribution channels. Describe whether you intend to use a direct sales force, indirect distribution channels, or combination of the two. Show detailed analysis for each choice, such as expected sales revenue, cost of sales (e.g. discounts, returns, promotions), and break-even point for each option. Be realistic about incentives required by the distribution channels to favor you over the competition. Also refer to Lecture Note Seventeen for a similar plan.

Business Plan – Sales and Distribution Section
- Sales and distribution channels:
    - How do you price your product?
    - Will you sell direct or indirect?
    - What is the cost of sales?
    - Show break-even analysis.
    - Why should channel members sell your product over competitors?

H. Financial Statements

In this section, show the numbers reflecting your projections for sales revenues and expenditures. Using spreadsheet software, show complete financial statements for the next three to five years. These include: balance sheet, income statement, and the cash flow statement. You can find a detailed listing of the financial statements in any accounting book. Also show a detailed sensitivity analysis, which typically includes three scenarios: three to five year projections for expected sales and expenditures, as well as anticipating for results that are below and above the expected target sales and expenditures. Services of accountants are essential to get you started in the right direction.

In addition, show detailed costs associated with achieving each milestone. Determine the type of financing you intend to secure, such as borrowing money from banks, or by venture capitalists investing in your company.

Business Plan – Financial Statements Section
- Financials:
    - Show three to five years of spreadsheets for:
        - Income Statement.
        - Balance sheet.
        - Cash flow.
    - Show sensitivity analysis, which is the same as "what if analysis".
    - The cost of achieving each milestone.
    - How many rounds of financing will you need?
    - Why invest in your company versus other start-ups?

I. Exit Strategy

A good business plan also incorporates an exit strategy, meaning even if your business is successful, how will you payback the initial investors and founders. There are a few options available to choose from: Initial Public Offering (IPO), selling out or getting acquired by another company, and merging with another company. There are advantages and disadvantages associated with each option, which requires a very careful examination of each. In recent years, when the stock market has been extraordinarily high, the IPO has been a favorite option for successful start-up companies.

Business Plan – Exit Strategy Section
- Exit strategy/liquidity:
    - Initial Public Offering (IPO).
    - Selling out.
    - Merger.
    - Getting acquired.

J. Appendix

Include succinct, relevant and credible supporting documents in this part of the business plan. Detailed information about technology and market research are also provided in this section. In addition, any relevant miscellaneous information should be included.

IV. Additional Business Plan Outlines

There are many reference sources (such as books, software programs, and web pages) to help you write a business plan. Some of the web sources are listed below:

Business Plan – Additional Outlines
- (http://www.bplans.com/)
    - This site has many samples.
    - Reviews business plan samples for high-tech start-ups.

- (http://www.sba.gov/starting/indexbusplans.html)
    - This is the Small Business Administrations (SBA) site, which is a part of the US government.
    - It has information about business plan outlines.

- Also check out: (http://www.sba.gov/starting/indexstartup.html).

V. Operating (Implementation) A Business Plan

A business plan articulates the overall strategy and tactics for a business. It also describes the high-level milestones and resources needed to accomplish these goals in a timely manner. However, the main complexity is in tactical implementation, making execution the key to success for a winning business plan. Successful execution of tactics is always more important than having a brilliant strategy.

Among the most difficult resources to manage are the human resources, the very people who must implement our business plan. That is, finding talented people and managing creative teams by keeping them focused and motivated, especially after a disappointing result, as well as managing internal politics.

Keeping a project on its ***critical path*** (i.e. interdependent tasks, which can not be delivered late) is crucial to a successful implementation plan. To do so, you must create a detailed flowchart, and accurately define interdependent tasks, headcount, and cash needed to execute the plan. There are many commercially available project management software tools, which you can utilize when developing a comprehensive implementation plan.

Implementation Plan
- Execution, execution, execution.
- This is the tactical implementation of your strategy.
- A great execution is always better than a great strategy.
- Focus on:
    - Detailed flowchart for logistics.
    - Resources in general; human resources in particular.
    - Critical path and interdependent tasks.
    - Contingencies, change course if necessary.
    - High or low demand, how will it affect you.

You should be flexible enough to modify your business and operating plans. Changing a business plan is very common in start-ups, but usually very difficult to implement. Any change requires money and human resources to implement, and in most cases, introduces delays and cost overruns. In a simple case scenario, just imagine what would happen if the actual demand was higher or lower than you had planned. Either way, the cost will be higher. If the demand is higher than supply, you may lose future market share, which usually is very expensive to get back in the long run. If the demand is lower, carrying a large underutilized inventory is very expensive. Therefore, you need to think of contingency plans that can be easily and quickly implemented.

The key to the successful implementation of a business plan is communications. A successful entrepreneur must be a capable leader with great communication skills.

Final Thought

A successful entrepreneur learns business by doing. They maintain their technical knowledge by constantly learning new advances in technology. They are energetic communicators, great public speakers, and effective listeners. They find many mentors throughout their careers. They understand the value of having been associated with an extensive network of professional people. They know how to manage people and risks effectively.

Closure

We have covered a vast array of topics in this lecture, but this is only the starting point for engineers with a burning desire to become successful entrepreneurs. There are many books written on this subject and many public and private sources to get help from. Be sure to seek help when you need it.

Essential thoughts in this lecture

Issues.	Applicability to your project, if any.
What is the entrepreneurial concept?	Entrepreneurs cannot wait to begin their own business. They will consider all possibilities including school projects. Is this notion applicable to your group?
Do you want to add anything else?	Please elaborate.

LECTURE NINETEEN: *Entrepreneurial and Venture Capital (Continued)*

Remark: The purpose of this lecture is to introduce engineering students to entrepreneurial and new venture issues as well as financing resources. In this lecture, we discuss issues that an entrepreneur must consider when building a successful startup company as well as various financing choices.

The following lecture is a continuation of the previous lecture prepared and delivered at the University of California at Riverside by Mr. Kayvan Mozaffari – a former student of this instructor.

Introduction

Here, we discuss various sources of financing a high-tech startup company as well as the importance of finding the required professional help necessary to build a successful company. Getting the right advice from professionals such as accountants and attorneys is often the most important aspect in building a successful company. Also, choosing the right source of money to finance a high-tech startup is among the first crucial business decision that an entrepreneur must make. The materials covered herein are related to materials covered in Lecture Notes Eight, Nine and Eighteen.

I. Professional Help

Entrepreneurs understand and value the help of other professionals in order to build a lasting and successful business. They will need all kinds of help from a wide range of professionals from attorneys, accountants, scientific and technical advisors, to business advisors and board of Directors.

> **Who Can Help Entrepreneurs**
> - To build a successful business, the entrepreneur needs professional help from:
>   - Attorneys.
>   - Accountants.
>   - Scientific and technical advisors.
>   - Business advisors.
>   - Board of directors.
>   - Venture capitalists.

A. Attorneys

There are many practical and legal reasons for entrepreneurs to hire attorneys right from the start. Good attorneys have an extensive track record of working with successful companies in a specific industry. To find a good attorney, seek referrals from others in the industry and check their references carefully. There are many attorneys, but select those that are excellent in their field.

Attorneys will help entrepreneurs with setting up their business as a corporation or partnership, as well as providing introductions to investors, accountants, and other good attorneys with different specialties.

Attorneys and Their Functions
- Make sure you hire only good attorneys.
- Good attorneys have a track record in your industry.
- Seek referrals; check their references; interview them.
- They will help you set up your company.
- They will stop you from making mistakes.
- They will extend your network.
- They will introduce you to investors, accountants, *etc.*.

B. Accountants

Accountants are an absolute necessity and will safeguard a business by helping entrepreneurs to understand how their business is performing. They highlight weak points and prevent entrepreneurs from making expensive mistakes, as well as avoiding legal troubles. They setup useful financial processes to assist the financial management of the business. In addition to an accountant, there is also a need for a reputable accounting firm in conducting an independent audit of the books annually.

Accountants and Their Functions
- A good accountant will:
  - Save your company.
  - Help you measure the true performance of your business.
  - Show you the weak points and trouble signs.
  - Will set up useful financial processes.
  - Stop you from making expensive mistakes.
  - Keep you out of legal troubles.
- You will need to hire a reputable accounting firm to audit your books each year.

C. Choosing Scientific/Technical and Business Advisory Board Members

Members of the advisory board are an entrepreneur's best source to locate great mentors. Combined, they have the experience, which an entrepreneur needs and may not have. They have current, relevant information and know people (good and bad), and can help when needed. Typically, an advisory board consists of five to ten respected people from the business, industry, and academia. Close friends and relatives should be avoided. Entrepreneurs should regularly speak with their advisory board members and reimburse them for their consultation (in the form of equity and money), and expenses if any.

Scientific, Technical and Business Advisory Board Members and Their Functions
- Typically you need five to ten people.
- Knowledgeable and respected in business, industry, and academia.
- They are your mentors in different areas.
- They have the experience that you do not.
- They have up-to-date information in their area of expertise.
- They know many people in your field.
- Do not pick relatives and friends.
- Give the above members equity for their contributions.

D. Choosing Board of Directors

The board of directors usually consists of five to seven people, which represent the shareholders interests. Their primary focus is to maximize shareholders' value, and help with setting-up company strategy. Also, they approve budgets and business plans. They check on management, and interview and hire top level executives. They are experienced business people (such as venture capitalists) and get paid sometimes in the form of equity. Since it is difficult to remove them from the board, entrepreneurs must be extremely selective in choosing them.

Board of Directors and Their Functions
- Usually you will need five to seven people.
- They represent the shareholders interests.
- Help with setting company strategy.
- Approve budgets and business plans.
- Check on management.
- Interview and hire top level executives.
- They are experienced business people (such as venture capitalists).
- They get paid sometimes in the form of equity.

II. Financing a Venture

There are many ways to finance a start-up venture: Equity versus debt, and debt versus bootstrapping (i.e. entrepreneurs use their own funds), which depends upon an entrepreneur's particular situation and experience. Entrepreneurs must make several crucial decisions regarding investment amount, type, and attractiveness. How much control will they maintain (solicited or unsolicited help comes with investments)? What are the pay back requirements? These decisions should primarily revolve around the confidence that entrepreneurs have in the success of their idea within the specified time frame. It is more important to be realistic than optimistic.

> **Financing Your Venture**
> - You can choose any combination of:
>   - Equity.
>   - Debt.
>   - Bootstrapping.
> - Must decide on:
>   - Amount, type, attractiveness.
>   - Level of control you want to let go.
>   - Pay back requirements.

A. Equity Financing

Equity financing means raising cash in exchange for ownership interest (of stocks) in a company.

1. Venture Capital

Venture capitalists have access to money sources that are allocated to high risk and high return investments. Typically, they have access to a pool of funds over $100-$200 million to invest in a handful of startup companies in less than five years. Their salary comes from management fees and takes up to a third of the profits. An experienced and successful venture capitalist that has been around for a long time understands it is not that simple to make money consistently. Not every deal will bring a desired return, and in many cases, they will lose money.

> **Venture Capital Financing**
> - Equity Financing - Venture Capital:
>   - Have access to large amount of funds.
>   - They are looking for high risk/reward.
>   - Pool of funds to invest in handful of start-ups.
>   - They take salaries and part of the overall returns.
>   - Good ones have been around for a while.
>   - They look for 10 to 20 times of return, in order to cover for non-performing start-ups.

The venture capitalists must have a great network of relationships in the industry. They know many investors, technologists, academics, industry leaders, and have access to a pool of technical, qualified and energetic people, bankers, attorneys, accountants, and entrepreneurs. When needed, entrepreneurs have access to this network.

Venture Capitalists

- Venture capitalists MUST have a great network in your industry:
    - Investment Bankers.
    - Technologists, academia, industry leaders.
    - Engineers, medium and high level managers, marketers, business developers.
    - Attorneys, accountants, entrepreneurs.

The venture capitalists consistently receive (and may read) well over 200 business plans a year and normally choose less than a handful to monitor, support, and fund (usually over $1 million). The leading venture capitalist is responsible to do all detail work such as multiple meetings with entrepreneurs, conduct the "due diligence" process by contacting the industry leaders, academics, target customers, and checking references. If there is high potential for a large market and they feel that they can actually work with the entrepreneur, they will take the deal to their partners. If the deal gets approval from other partners, the "term sheet" is prepared, which articulates the monetary terms of the deal. After the negotiation period, attorneys will get together to close the deal. However, entrepreneurs should only count on the money when it is actually in their bank account.

At this point, the venture capitalist will have a seat on the Board of Directors. They will help entrepreneurs to set goals and milestones, review their progress, keep their companies focused, and hire executives (CEO, CFO…). But most importantly, they monitor the cash expenditure rate and raise additional money. They help entrepreneurs and guide them in making difficult decisions (for example, going public, merging, getting acquired, laying off and shutting down) when it is necessary.

Venture Capitalists (Continued)

- They receive several hundreds business plans per year.
- They only fund a handful, typically over $1 million each.
- Each deal goes through many screening levels.
- They usually have a seat on the Board of Directors.
- They are there to help and monitor.

How would an entrepreneur know if he/she is working with a good venture capitalist? Good venture capitalists have a successful track record and a large number of contacts in all areas, particularly in the same industry as the entrepreneur. They are expected to resolve very difficult situations, make time for the entrepreneur, respect them, listen to their business problems, as well as help them to find the right solution. They are well informed and have a very good understanding (both technical and business) about the entrepreneur's industry. They are interested in helping the entrepreneur to build a lasting company, not just interested in "going public" or selling out, so they can cash in quickly. Typically, they want a 10 to 20 times of return in less than three to five years, depending on the market. They also have their own internal allocations, as well as formulas for company valuations.

2. Angels

Who are they and what do they do? Angels are mostly ex-entrepreneurs that have already made a significant amount of money and want to repeat the process. They typically get involved before venture capitalists. They pay more and ask fewer questions, but they have lower expectations than venture capitalists. They are usually well connected and have a very good reputation in the entrepreneur's industry with extensive technical knowledge. The best angels are usually former entrepreneurs with strong and extensive venture capitalist connections. They provide the money, but they are usually too busy to help entrepreneurs. Also, beware of shaky, inexperienced, and non-reputable angels.

3. Bootstrapping

Bootstrapping is defined as the process by which an entrepreneur uses his/her own money to finance his/her company. Unless an entrepreneur is very wealthy, it is not recommended. Bootstrapping is not a solid financing strategy. It would be very difficult, if not impossible, to bootstrap a high growth business. The chances of cash shortages and failure are very high, which may expose an entrepreneur. In this instance competitors may benefit, or the entrepreneur may even give away their ideas completely.

Angel Financing and Bootstrapping

- Angels:
    - Usually ex-entrepreneurs and wealthy individuals.
    - Usually get in before venture capitalists.
    - Pay more and want less return than venture capitalists.
    - Should not expect to receive management help from them.
- Bootstrapping:
    - Not recommended unless exceptional cases.
    - Chances of cash shortages and failure are very high.

B. Debt Financing

Debt financing is a loan agreement to borrow cash and repay the full amount plus interest within an agreed time period.

1. Bank Loans

Bankers have no dilutions of shareholders equity, and entrepreneurs are obligated to pay back principal plus interest. Bank loans usually work best when there are fixed assets as collaterals. However, lenders do not help entrepreneurs with management of the business. Entrepreneurs should consider this purely as a source of cash.

2. Loans from Relatives

Entrepreneurs will have to live with relatives, so they must choose wisely. Relatives are usually not in the deal for the return, so entrepreneurs should do their best to pay them back.

3. Loans from Suppliers and Customers

Suppliers and customers are a great source of securing inexpensive capital. These loans are in the form of equipment, raw materials, and (sometimes) cash from suppliers, similarly, purchase orders and cash advances from customers. They have an incentive to see entrepreneurs succeed and will help them to do so.

Debt Financing

- Debt Financing
    - Bank loans:
        - No dilution of shareholders equity.
        - Usually works best when fixed assets are involved.
        - Not qualified to help.
    - Loans from relatives and friends:
        - Are they really serious?
    - Loan from suppliers and customers:
        - A great source of help when you need it.
        - A better source than family and friends.

III. Additional Information Sources

Before committing to any of the above, do additional research using the following sources.

Information Sources About Venture Capitalists Activities

- Price Waterhouse Venture Capital Survey is available in:
    - www.pw.com
- San Jose Mercury – Quarterly Venture Capitalists funding report is available through business/moneytree in:
    - www0.mercurycenter.com
- Red Herring is a magazine focused on start-ups and financing activities:
    - www.herring.com
- Monitor the active organizations in your local area.

IV. Growing your company

There are two sources of funds to grow a business: (1) Internal – cash is generated by sales revenue. (2) External – cash is generated from debt (borrowing cash to repay) or equity (raising cash in exchange for ownership from private and public). Entrepreneurs can expand their business by building it themselves (also called organic growth), or expand through acquisitions (buy other companies). There are many advantages and disadvantages associated with both strategies.

The advantage of organic and internal growth is that entrepreneurs have a good chance to build their company the way they like, having total control. A disadvantage is that it takes time to hire and manage a team in order to become effective and generate profit.

On the other hand, when an entrepreneur acquires a company, they save time and eliminate a competitor. They have immediate access to the company's customer base, and therefore, an immediate jump in revenues. Also, they have access to functioning teams, resources, and additional products and intellectual properties to offer to their existing customer base. A disadvantage is that they are in charge of making this acquisition a success. Not an easy task, they could end up losing a lot of money and be forced to divest. Typically, entrepreneurs need to raise a substantial amount of cash from venture capitalists, investment bankers, and banks, which are specialized in mergers and acquisitions (M&A).

V. Initial Public Offering – IPO

The advantage of going public is that the entrepreneur may cash-in along with other people who have helped to grow the company. But now they must conform to the regulations of Security and Exchange Commission (SEC), and do the ongoing paperwork and filing. From this point on, the company lives in a glass jar; everybody can see their every strategic move, especially their competitors.

Final Thought

There are many facets to building a high-tech startup company. A successful entrepreneur understands the value of building a strong network of professional contacts and coaches. This is as important as being tenacious and technically brilliant. It is also important to recognize the value of teamwork in building a successful and growing high-tech company. Finally, it is crucial for entrepreneurs to make the right decisions about financing their startups, as well as hiring knowledgeable and motivated professionals.

Closure

We have covered a vast array of subjects in this lecture, but this is only the starting point. There are many books written on this general area, which are suggested readings for students with an interest in becoming an entrepreneur. More importantly, entrepreneurs learn during the entrepreneurial process and ask for help when they need. Also, it is important that they attend professional events and meet new people with diverse competencies.

We should explicitly state that although the titles of Lecture Notes Eight, Nine, as well as Eighteen and Nineteen imply "high-tech" products, there are plenty every day products that when improved may dominate the market for this product and its production becomes very profitable. For instance, a 20 cents plastic used in an engine of a $40, 000 car must be functional in order that the car runs. To manufacture that 20 cents plastic one has the same business concern as in any company. Thus, do not be intimidated by the so-called "high-tech" products – the materials suggested here are equally important and applicable to all kinds of products and their related businesses.

Not every company starts big and needs all the preceding expertise right away. Again, do not be intimidated by the elaborate organizational structure, which has been introduced herein, as a model for a very successful company. This information is a point of reference, and as stated before and throughout this textbook, one needs to exercise his/her good judgement at every step of the way. You must know the ultimate organizational structure, but it will take time to bring all of that in place.

Finally, to close our discussions on business aspects of this course we offer the following. When you have a strong feeling for a product and you really want to manufacture that profitably, explore all possible global opportunities for a potential saving as well as profitability. It always pays to look around.

Essential thoughts in this lecture

Issues.	Applicability to your project, if any.
How to build a high-tech startup company?	To build and deliver any high-tech product, one must understand the value of teamwork and having access to an extensive network of professionals.
Do you want to add anything else?	Please elaborate.

Acknowledgement:

This instructor is most grateful to and proud of his former student Kayvan Mozaffari for his contributions to this lecture note series. *He is also looking forward to benefit from the future expertise of his current and future students in improving and expanding this lecture note series for future editions hopefully.*

LECTURE TWENTY: *Quality Control*

Remark: In this lecture preliminary issues pertaining to the international system of standards and quality control in an engineering project are discussed.

The author is grateful to Dr. Farrokh Nassirpour – his former colleague at the University of Wisconsin-Madison, for developing the major part of this lecture. As well as Professor Harold J. Steudel, also from the University of Wisconsin-Madison, for generously providing the author with substantial materials from his teaching notes on these topics [4].

References:

[1] W. E. Deming, *Out of the Crisis.* Cambridge MA: MIT, Center for Advanced Engineering Study, 1986.
[2] J.M. Juran, Editor-in-Chief, *Juran's Quality Control Handbook.* New York: McGraw-Hill, 4th Edition, 1988.
[3] Quality System (QS-9000). Southfield, MI: Automotive Industry Action Group (AIAG), 3rd Edition, 1998.
[4] H.J. Steudel, *Various Classroom Lecture Notes on Quality and Quality Management.* Dept. of Industrial Engineering, The University of Wisconsin-Madison, 2001.

Introduction

Generally speaking, the importance of and the awareness about *quality*, or *quality control* as is referred to in the literature, has experienced a remarkable growth in recent decades. The word quality is used extensively to describe a host of issues that ultimately refer to various sub-areas of this general topic, which has been known for a long time as simply quality control. These sub-areas are too numerous to count herein, for instance, *quality management* and *total quality* are two such examples. Thus, the reader should realize that we are looking at the most generic aspects of this topic, in order to describe its most fundamental issues that serve as the basic knowledge to complement a well-rounded general engineering education that we hope to achieve.

It is a well-known fact that an industrial society provides its citizens with an array of amazing benefits from technology. However, the utilization of such benefits makes the continuity of the corresponding life style dependent on the quality of the goods and services provided to the community, which these activities in turn form the bases for an industrial society. The international competition in quality has become very intense. Now, the quality is a critical element in every facet of human life, in the forms of international trade, defense, human safety, health, and maintaining environment.

The concept and application of quality as professionals know it, requires that extensive literature be consulted, and to cover those topics in one lecture is a challenge. In order to make sense, however, we have focused on the *automotive industry*, and cover briefly the worldwide status of quality in that industry. One should realize that the automotive industry is the pioneer in the field of quality and has always been the leader in development and application of quality "tools" and "ideas". The basic motivation for the extensive application of quality in most industries worldwide is the intense competition faced by those industries.

In response to the worldwide pressure, the automotive industry started to implement quality concepts internally, and by publishing sets of standards and certification systems, they also forced their suppliers to follow suit. In recent years the automotive industries have adopted the international standards and have joined together to publish new standards utilizing an organization called Automotive Industry Action Group

(AIAG). Before proceeding further, two points are important to remember.

1. Major foundations of quality concepts and tools are probability, statistics, and their applications in manufacturing and services.
2. The foundation of International Standards Organization (ISO 9000) has been developed in the United States by the automotive and defense industries.

Therefore, this is the reason that the new engineering programs require at least one course in applied statistics for all its disciplines. With such a prerequisite we can continue a formal and rigorous study of this topic. In this lecture, however, we study only a general and conceptual review of quality and related topics. Thus if we wish to give a road map for such a broad study, then we propose to give a generic definition for a system of standards followed by at least one way of attaining and sustaining that system.

In the most general sense, a standard that qualifies and quantifies the quality of a product is only a point of reference, i.e., we have a definition here. Therefore, there are degrees of reaching that point of reference, i.e., we need measurement and thus be concerned with its accuracy and reliability. Finally, we need to keep the operation at or around certain conditions, in order to attain this point of reference and sustain the production, i.e., we must continuously learn ways of attaining and managing a process for high quality outcomes. Having said the above, we cover the corresponding system of international standards, as much as possible, in our allotted time: and followed by one method of attaining such high quality performance, as a point of view, to improve the quality and thus to achieve, as closely as possible, the standard.

I. Basic Definitions

The word quality has several meanings. According to Juran's Quality Control Handbook, two of those meanings are [2]:

1. "Quality consists of those product features which meet the needs of customers and thereby provide product satisfaction."
2. "Quality consists of freedom from deficiencies."

Before proceeding further, we recall that the word ***continuously*** means without interruptions, for example a river flows continuously, while the word ***continually*** means with regular interruptions or intermittent. For example, waves from the sea beat continually. Occasionally and in our own optimism we mix these words, for instance, we often say that we are continuously learning, while we forget that that is impossible since we sleep a few hours every day.

II. Quality System Requirements (QS-9000)

Quality system requirements (QS-9000), and the other 11 related manuals published by AIAG [3], cover most quality activities in automotive industries in the world. These manuals are about the following topics:

- Quality system assessment (QSA)
- Measurement system analysis reference manual (MSA)

- Statistical system analysis reference manual (SPC)
- Potential failure mode and effect analysis reference manual (FMEA)
- Production part approval process (PPAP)
- Advanced product quality planning and control plan reference manual (APQP)
- Semiconductor supplement
- AEC-A100: QSA semiconductor edition
- Tooling and equipment supplement
- Tooling and equipment QSA-TE
- Reliability and machinability guidelines for manufacturing machinery and equipment.

Certainly, we cannot give detail of all these matters herein. However, the following brief review should help to elucidate the main points covered in some of these manuals. Here, we give briefly a series of concepts and definition to familiarize the reader with the main issues in quality.

The QS-9000 refers to the entire standards, which has two main sections: Section one is the ISO-9000 based requirements; and Section two is the customer-specific requirements. *The goal for QS-9000 is the development of fundamental quality systems that provide for the continuous improvement, emphasizing defect prevention and the reduction of variation and waste in the supply chain.* In this section, we cover the main ideas of QS-9000.

A. Management Responsibility

One of the main responsibilities of management is to define and document its policy for quality, including objectives for quality and its commitment to quality. The quality policy shall be relevant to the supplier's organizational goals and the expectation and need of its customers. The internal and external suppliers shall ensure that this policy is understood, implemented and maintained at all levels in the organization.

The supplier shall utilize a formal, documented and comprehensive business plan. The business plan shall be a controlled document. The content of business plan is not subject to a third party audit. This plan typically include as applicable:

- Market-related issues
- Financial planning and cost
- Growth projections
- Plant/facilities plans
- Cost objectives
- Human resource development
- R & D plans, projections, and projects with appropriate funding
- Projected sales figures
- Quality objectives
- Customer satisfaction plan
- Key internal quality and operational performance measurables
- Health, safety and environmental issues.

B. Quality System Elements

The supplier shall establish, document and maintain a quality system as a means of ensuring that product conforms to specified requirements. The supplier shall prepare a quality manual covering the requirements of the ISO standards. The quality manual shall include or make reference to the quality system procedures and outline the structure of the documentation used in the quality system.

C. Continuous Improvement

The supplier shall continuously improve in quality, service (including timing, delivery), and price that benefit all customers. This requirement does not replace the need for innovative improvements. In this regard, we elaborate further in Section VIII, using materials from [4].

D. Design and Development Planning

The supplier shall prepare plans for each design and development activity. The plan shall describe or reference these activities, and define responsibility for their implementation. The design and development activities shall be assigned to qualified personnel equipped with adequate resources. The plans shall be updated as the design evolves.

The supplier's design activity shall be qualified in the following skills as appropriate:

- Geometric dimensioning and tolerancing (GD&T)
- Quality function development (QFD)
- Design for manufacturing (DFM) and design for assembly (DFA)
- Value engineering (VE)
- Design of experiment (DOE)
- Failure mode and effect analysis (FMEA)
- Finite element analysis (FEA)
- Solid modeling
- Simulation techniques
- Computer-aided design (CAD)/Computer-aided engineering (CAE)
- Reliability engineering plans.

E. Document and Data Control

The supplier shall establish and maintain documented procedures to control all documents and data that relate to the requirements of this international standard including, to the extent applicable, documents of external origin such as standards and customer drawings. The example of appropriate documents includes:

- Engineering drawings
- Engineering standards
- Math (CAD) data
- Inspection instructions
- Test procedures
- Work instructions
- Operations sheets
- Quality manual

- Operational procedures
- Quality assurance procedures
- Material specifications.

F. Process Monitoring and Operator Instructions

The supplier shall prepare documented process monitoring and operator instructions for all employees having responsibilities for operation of various processes. These instructions shall be accessible at the work station. Process monitoring and operator instructions shall include or reference, as appropriate:

- Operation name and number keyed to the process flow diagram
- Part name and number, or part family
- Current engineering level/date
- Required tools, gages and other equipment
- Material identification and disposition instructions
- Customer and supplier designated special characteristics
- SPC requirements
- Relevant engineering and manufacturing standards
- Inspection and test instructions
- Reaction plan
- Revision date and approvals
- Visual aids
- Tool change intervals and setup instructions.

III. Advanced Product Quality Planning and Control Plan

Product quality planning is a structured method of defining and establishing the steps necessary to assure that a product satisfies the customer. The goal of product quality planning is to facilitate communication with everyone involved to assure that all required steps are completed on time. Effective product quality planning depends on a company's top management commitment to the effort required in achieving customer satisfaction. Some of the benefits of product quality planning are:

- To direct resources to satisfy the customer
- To promote early identification of required changes
- To avoid late changes
- To provide a quality product on time at the lowest cost.

The product quality planning timing chart (as depicted in Fig. 2) and the product quality planning cycle (as shown in Fig. 3) require a planning team to concentrate its effort on defect prevention.

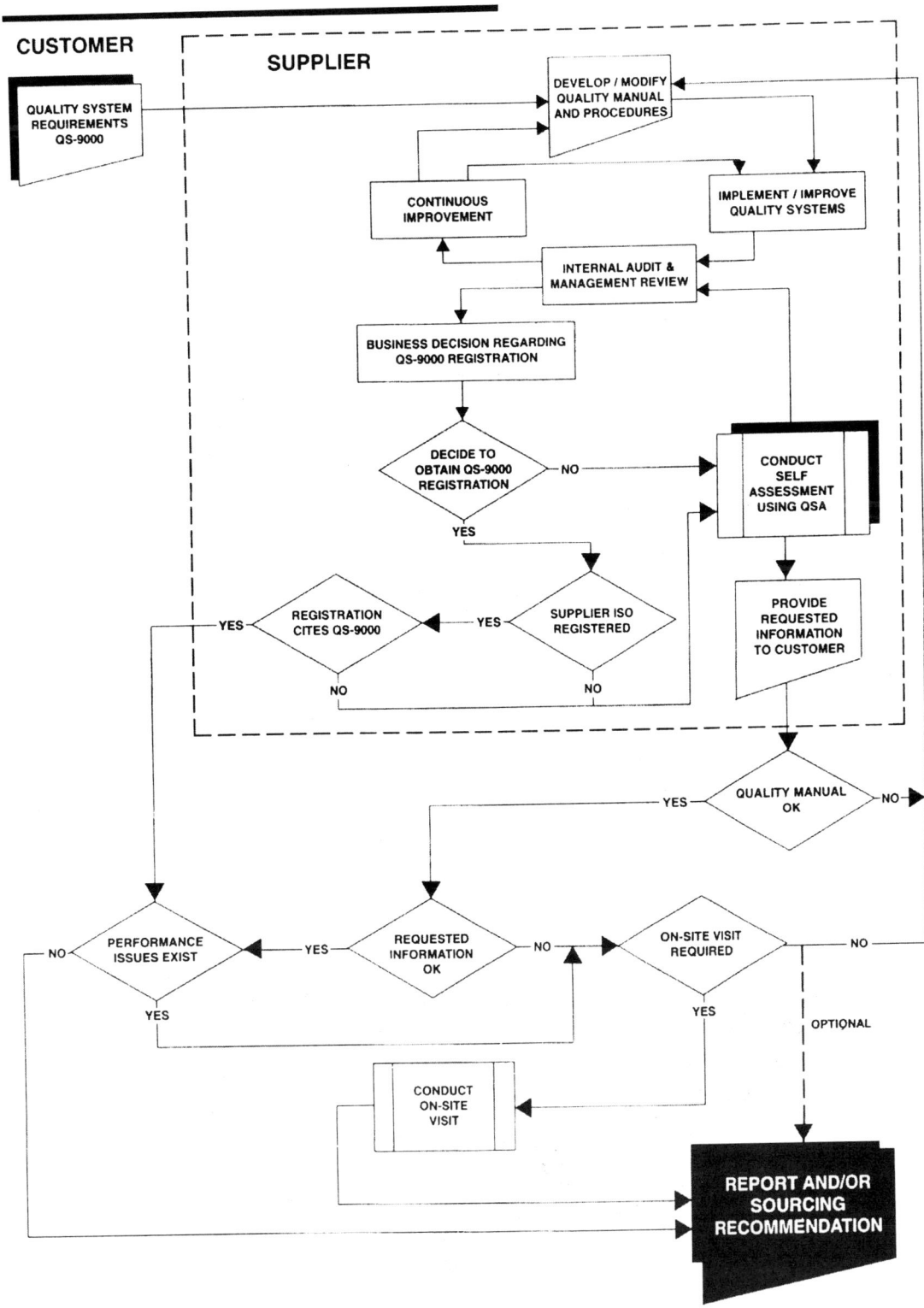

Fig. 1. The QS-9000 process.

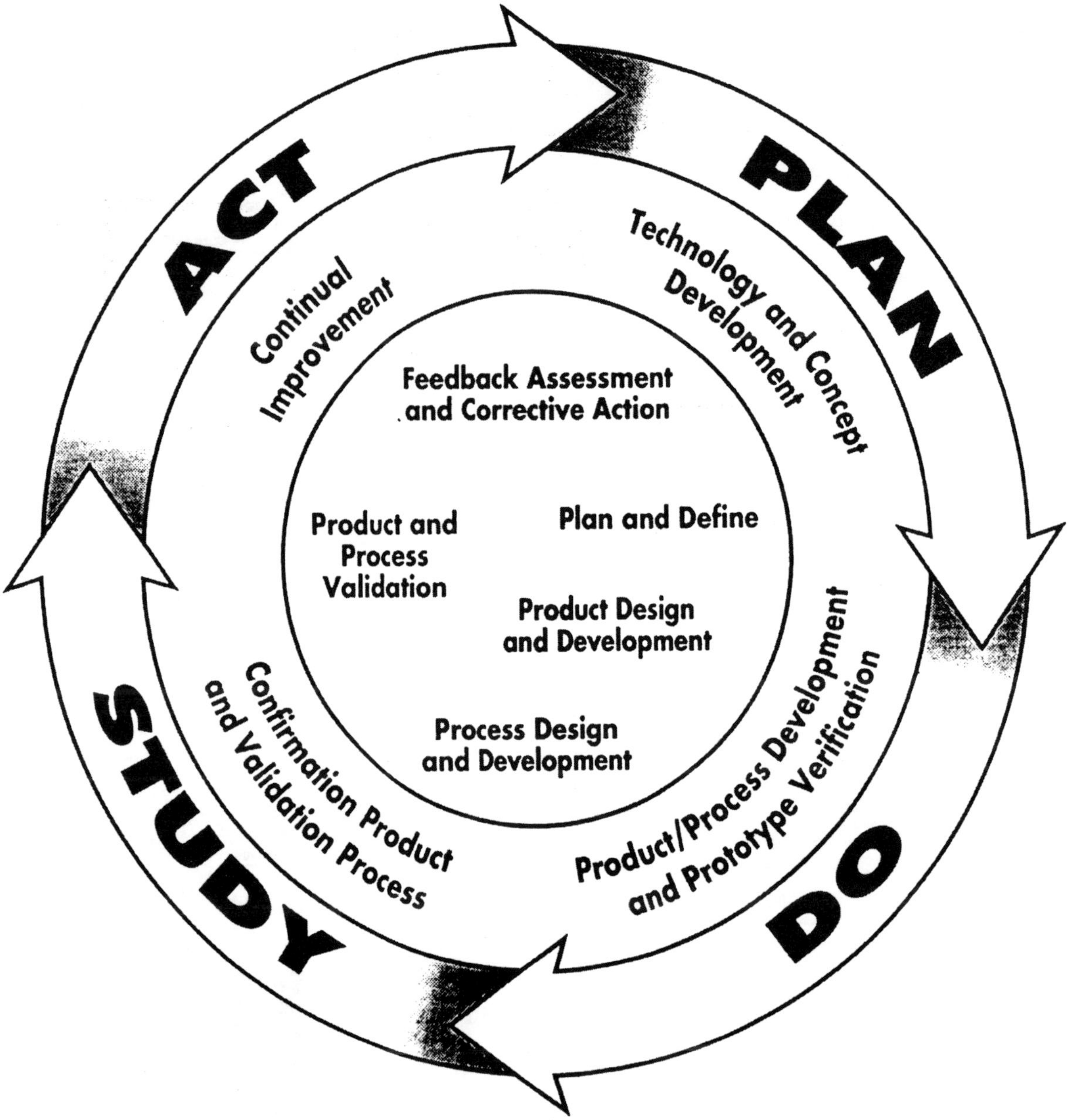

Fig. 2. Product quality planning cycle.

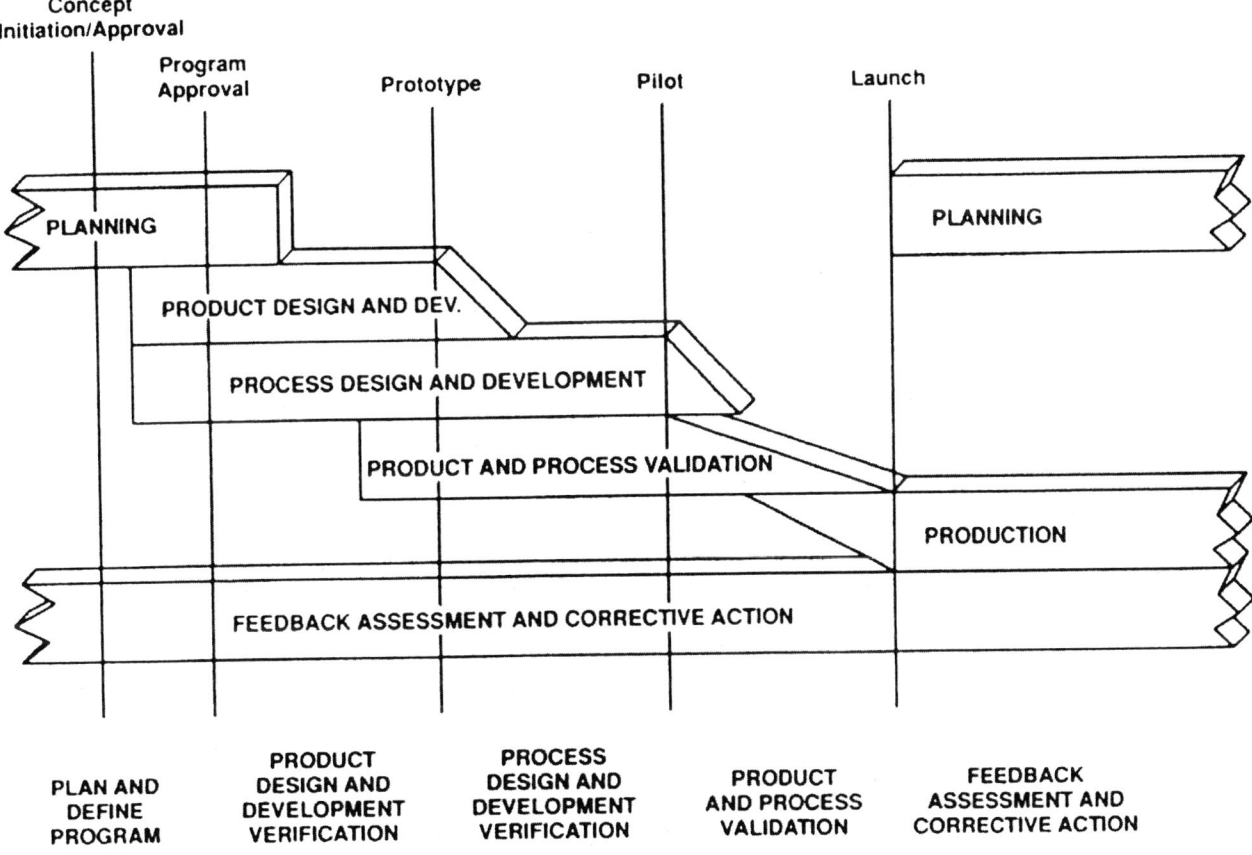

Fig. 3. Product quality planning timing chart.

A. Control Plan Methodology

The purpose of this control plan methodology is to aid in the manufacture of quality products according to customer requirements. It does this task by providing a structured approach for the design, selection and implementation of value-added control methods for the total system. Control plan provides a written summary description of the systems used in minimizing process and product variation.

B. Analytical Techniques

Some of the analytical tools used in the advanced product quality planning and control plan are as follows:

- Assembly build variation analysis
- Benchmarking
- Cause and effect diagram
- Characteristics matrix
- Critical path method
- Design of experiments (DOE)
- Design for manufacturing and assembly
- Design verification plan and reports (DVP&R)
- Dimensional control plan (DCP)
- Dynamic control plan (DCP)
- Process flow charting
- Quality function deployment (QFD)
- System failure mode and effects analysis (SFMEA).

Fig. 4. Control plan.

IV. Production Part Approval Process (PPAP)

Production part approval process (PPAP) defines generic requirements for production part approval, including production and bulk materials. The purpose of PPAP is to determine if all customer engineering design record and specification requirements are properly understood by the supplier and that the process has the potential to produce product consistently, meeting these requirements during an actual production run, and at the quoted production rate. The major elements of PPAP are:

- PPAP process requirements
- Customer notification and submission requirements
- Submission to customer
- Part submission status
- Record retention.

V. Potential Failure Mode and Effects Analysis (FMEA)

The FMEA can be described as a systematic group of activities intended to: (a) recognize and evaluate the potential failure of a product/process and the effects of that failure, (b) identify actions that could eliminate or reduce the chance of the potential failure occurring, and (c) document the entire process. It is complementary to the process of defining what a design or process must do to satisfy the customer. Every FMEA focuses on the design, whether it is of a product, or a process. There are two types of FMEA's: (1) Design FMEA, and (2) Process FMEA.

VI. Measurement Systems Analysis (MSA)

Measurement data are used more often and in more ways than ever before. For instance, the decision to adjust, or not, a manufacturing process is now commonly based on measurement data. Measurement data, or some statistics calculated from them, are compared with statistical control limits for the process, and if the comparison indicates that the process is out of statistical control, then an adjustment of some kind is made. Otherwise, the process is allowed to run without adjustment.

Another use of measurement data is to determine if a significant relationship exists between two or more variables. For instance, it may be suspected that a critical dimension on a molded plastic part is related to the temperature of the feed material. That possible relationship could be studied by using a statistical procedure called regression analysis to compare measurements of the critical dimension with measurements of the temperature of the feed material. Studies that explore such relationship are examples of what Deming called analytic studies [1]. In general, an analytic study is one that increases knowledge about the system of causes that affect the process. Analytic studies are among the most important applications of measurement data because they lead ultimately to better understanding of process. The benefit of using a data-based procedure is largely determined by the quality of the measurement data used. If the quality of data is low, the benefit of the procedure is likely to be low. Similarly, if the quality of the data is high, the benefit is likely to be high as well.

The quality of measurement data is related to the statistical properties of multiple measurements obtained from a measurement system operating under stable conditions. For instance, suppose that a measurement system, operating under stable conditions, is used to obtain several measurements of a certain characteristic in a process. If the measurements are all "close" to the master value for these characteristics, then the quality of the data is said to be "high." Similarly, if some, or all, of these characteristics are "far" away from the master value, then the quality of the data is said to be "low."

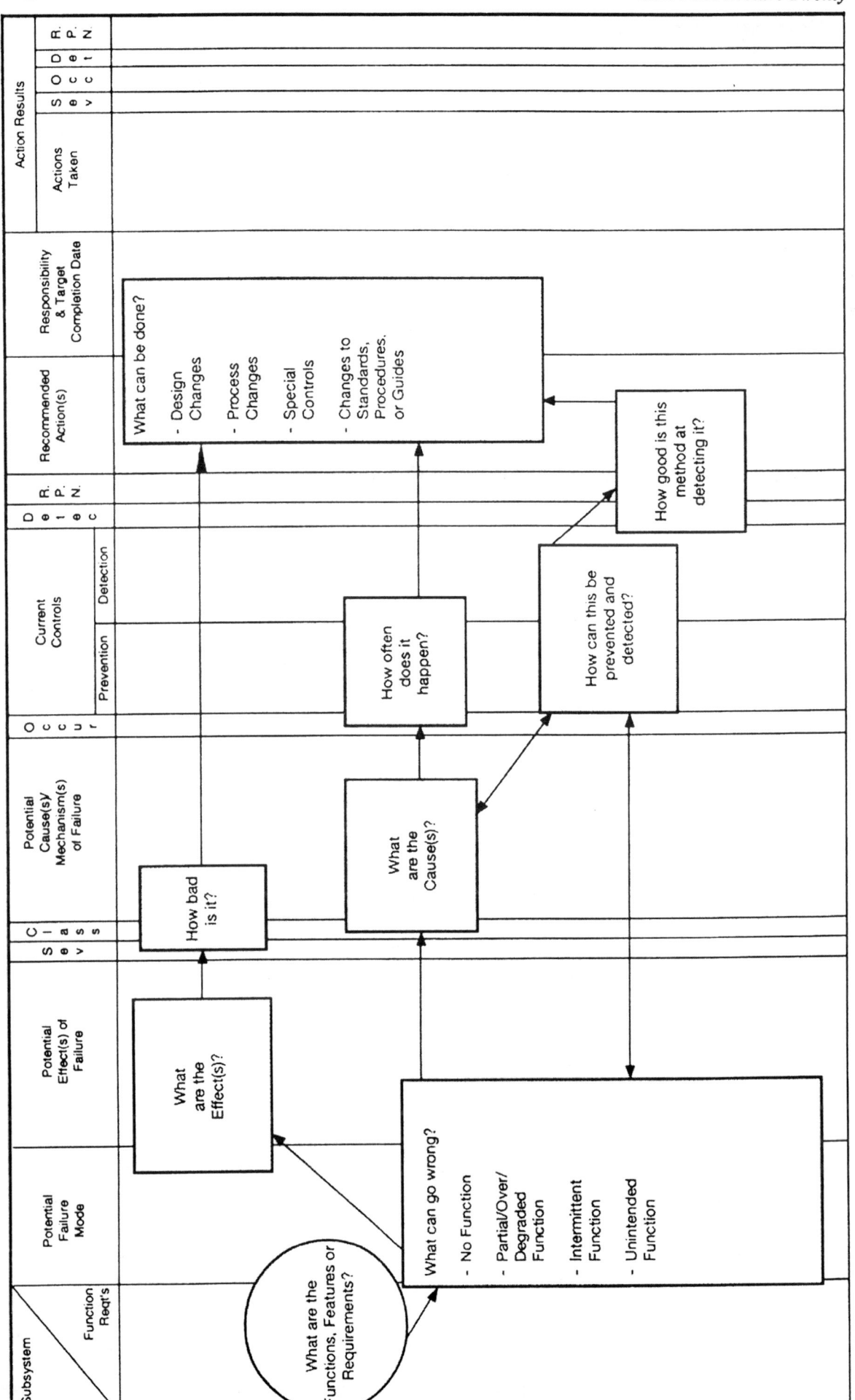

Fig. 5. FMEA process sequence.

The statistical properties most commonly used to characterize the quality of data are bias and variance. The property called bias refers to the location of a datum relative to a master value, and the property called variance refers to the spread of the data. But other statistical properties, such as the rate of misclassification, may also be appropriate in some cases.

One of the most common reasons for the low-quality data is having many changes in the data. For instance, a measurement system used to measure the volume of liquid in a tank may be sensitive to the ambient temperature of the environment in which it is used. In that case, variations in the data may be due either to changes in the volume or to changes in the ambient temperature. That makes interpreting the data more difficult and the measurement system, therefore, less desirable.

Generally, variations in a set of measurements are due to the interaction between the measurement system and environment. If the interaction generates too many variations, then the quality of the data may be so low that the data are not useful. For example, a measurement system with a large amount of variations may not be appropriate for use in analyzing a manufacturing process, because the variation in system measurement may mask the variation in the manufacturing process.

To manage a measurement system, we need to monitor and control the corresponding variation. This fact means that emphasis must be placed on learning how the measurement system interacts with its environment so that only data of acceptable quality are generated, since most other variations are undesirable. But there are some important exceptions. For instance, if the variation is due to small changes in the characteristics being measured, then it is usually considered desirable.

Procedures for assessing measurement systems assess the following statistical properties: repeatability, reproducibility, bias, stability, and linearity. Collectively, the procedures are sometimes referred to as "Gage R&R" procedures because they are frequently used only to assess the statistical properties of reproducibility and repeatability.

A. Statistical Properties of Measurement Systems

Although each measurement system may be required to have different statistical properties, there are certain properties that all measurement systems must have. These include:

- The measurement system must be in statistical control. This means that the variation in the measurement system is due to common cause only and not due to special causes. This is called statistical stability.

- Variability of the measurement system must be small compared with the variability in the manufacturing process.

- Variability must be small compared with the specification limits.

- The increments of measurement must be small relative to the smaller of either the process variability or the specification limits. A common rule of thumb is for the increments to be no greater than one-tenth of the smaller of either the process variability or the specification limits.

- The statistical properties of the measurement system may change as the items being measured change. If so, the largest (worst) variation of the measurement system must be small relative to the smaller of either the process variation or the specification limits.

VII. Continual Improvement and Statistical Process Control (SPC)

To continue to prosper in today's economic climate, manufacturers, suppliers and marketing organizations must all be dedicated to a continual improvement. These entities must constantly seek more efficient ways to produce products and services. These products and services must continue to improve in value. We must focus upon our customers, both internal and external, and make customer satisfaction a primary business goal.

A preventive strategy requires an understanding of the elements of a statistical process control system. The elements can be viewed as answers to the following questions:

- What is meant by a process control system?
- How does variation affect process output?
- How can statistical techniques tell whether a problem is local in nature or involves broader systems?
- What is meant by a process being in statistical control? What is meant by a process being capable?
- What is a continual improvement cycle, and what part can process control play in it?
- What are control charts, and how are they used?
- What benefits can be expected from using control charts?

VIII. Continual Improvement – Kaizen Concepts and Strategies

The following materials are closely gathered from [4], in order to present one successful philosophy for running a company. The "kai" in *Kaizen* means to break apart or disassemble, and "zen" means to improve feverishly. We must do "kai" first. *Kaizen* is a management philosophy and strategy to run a company. The concept centers on gradual, unbending improvement, doing little things better, continuously setting and reaching goals, and thereby establishing ever-higher standards.

Continuous process refers to a process, which never gets the final point, but improves continuously relative to the past. For example, *learning* is a continuous process of gaining knowledge, insight, understanding, and wisdom, while education is one strategy to foster learning.

Kaizen solves problems by establishing a corporate *culture* in which everyone (from the top management to the hourly personnel) can freely admit problems and work together to prevent their reoccurrence. In other words, a *continuous team effort* is to compress time rapidly, thus to eliminate or greatly reduce the waste in order to increase customer satisfaction.

Kaizen is also a *customer-driven strategy* for continuous and rapid improvement in the processes that affect cost, quality, and scheduling (in terms of volume and delivery).

Kaizen is a way to improve *all aspects* of an organization, and is the overriding concept of a good management.

A. Key Elements of Kaizen

The *team work* is the main issue in this philosophy and consists of the following basic points:

- Use a people-focused, and not a technology-focused approach to achieve competitiveness.
- Emphasize teams to gain collective knowledge and energies.
- Foster cooperation, mutual respect and common goals.

The next element is ***communication*** that consists of the following basic points:

- Involve people in solutions.
- Create and foster a culture where people are free to make and admit mistakes.
- Listen first, and speak second.
- Foster non-blaming and non-judgmental views.

Elimination of waste follows the above element with the following basic points:

- Identify and eliminate anything that does not add *"value"* to the product for the *"customer"*.
- Practice the "process oriented thinking".

The final key element of ***Kaizen*** is ***continuous "small steps,"*** which consists of the following basic points:

- Pledge to satisfy your customers by improving your processes continually.
- Take small steps to learn and to improve, but do it!

B. Change

Self-assessment is the prerequisite to any change, which in turn change is necessary to meet competitions and thus generate profits to sustain a company. Customer requirements are continuously increasing, and unless we are willing to meet this challenge, we cannot sustain the profitability of the company. Rapid growth in new equipment and expensive technologies is often limited and constrained in terms of effectiveness in customer satisfaction. So, the change is necessary.

A typical organization that has reached its "comfort zone," is often plagued by established levels and practices of waste. While the competition is involved in an on-going improvement of its outcome. Unless, a company is willing to change and improve dramatically, the competition will catch up and surpass it – improve or perish. A world-class company needs to implement new policies and practices that foster gradual, incremental, unending, and continuous improvement in its "processes." This change is done by using a team or group approach, which continuously identifies and implements improvements in all processes. Here, all means from shop floor to business practices. The company may have to reorganize and manage all operations as "processes," and measure the process improvement. Or the company must empower the workforce, and provide people with the knowledge, skills, and resources needed to achieve ***Kaizen*** as a way of life.

C. Managing Change

Companies must manage change involving continuous improvement in its processes and practices. If the management of the company fails to manage change, then the management must be changed. Managing change results in cyclical changes. For instance,

- Change of ***thought*** makes one's **behavior** change.
- Change of ***behavior*** makes one's **habits** change.
- Change of ***habits*** makes one's ***personality*** change.
- Change of ***personality*** makes one's ***destiny*** change.

Types of change may either be:

- Abrupt Change – *Innovation*, or
- Gradual Change – *Kaizen*.

The above **abrupt change** or **innovation** typically depends on the large investment and big breakthroughs due to: Technology, Equipment, and Massive Change. There are pros and cons associated with this concept that we cannot really elaborate fully. For instance, if we have a "great innovative leap forward," then we may not have sufficient funds or only a select few people to carry on. The drawbacks are potentially numerous, such as spending large sums of money and not having the total company involved. On the other hand, the **gradual change** or **Kaizen** typically depends on little or no investment and on-going small improvements in current "standards." This approach also involves total people in the company.

Maintenance, is that aspect of managing change, which refers to activities directed to maintaining current technological, managerial, and standard operating procedures (SOP). This aspect establishes policies, rules, directives, and standard procedures. As well as it provides training and maintains discipline.

In closing, this section, we remind the reader that there are extensive publications on this and similar management methodologies, which improve and sustain the standard of performance in a given company. We are running out of time and leave the rest to those of you interested to pursue this matter.

Final Thought

In this lecture note the genesis of the concept of quality is described. However, the main issue of applicable standard and quality control in its specific sense and as far as your project is concerned remains open. One must find the proper set of standards in his/her discipline or project and find a method of adhering to that set of standards – not a trivial task.

Closure

The lecture ends with the following remarks. We expect that each final report to cover a corresponding discussion on its quality aspects pertaining to the underlying project. This is an open and on-going study and we entertain and indeed encourage other methodologies to be consulted in this process.

Essential thoughts in this lecture

Issues.	Applicability to your project, if any.
How would you consider a set of standards relevant to your project?	If so, please elaborate.
Do you want to add anything else?	Please elaborate.

LECTURE TWENTY ONE: *Intellectual Property Issues*

Remark: In this and the following lecture issues pertaining to applying for intellectual property (IP) certificates and preliminary steps for applying for patents are discussed. The purpose of this and the subsequent lecture note is to clarify these matters and give further information on what some of these terminologies mean and how to secure an IP certificate or to seek a patent. Clearly, it is not our intention to serve as an attorney nor do we plan to provide a plethora of forms and documents in this regard. However, the following disposition is a good start to compile necessary information for this purpose.

In preparation of these notes, the author has been fortunate to have access to the vast resources at the University of Wisconsin-Madison. This depository includes the Technical Reports Center at Kurt F. Wendt Library; the Wisconsin Alumni Research Foundation (WARF); and the legal offices of the University of Wisconsin-Madison. Special thanks are due Ms. Nancy Spitzer of the Wendt Library; Ms. Deanna L. Dietrich, Esq., of the College of Engineering; Dr. Steven Price of the University-Industry Relations; Ms. Elizabeth L. R. Donley, Esq., of WARF; and Ms. Kathleen S. Irwin, Esq., of the University Administration. The information in the following two lectures is prepared initially by studying and restructuring the materials presented by the above colleagues in several lectures at the University of Wisconsin-Madison. Ms. Nancy Spitzer has compiled the following references, and also she has generated all subsequent forms.

Special Resources:

[1] Wisconsin Alumni Research Foundation (www.wisc.edu/warf) (Permission required).
[2] US Patent and Trademark Office (http://www.uspto.gov/).
[3] Patent Search (http://www.uspto.gov/patft/index.html).
[4] U.S. Patent Classification (http://www.uspto.gov/go/classification/).
[5] Foreign Patents Search (http://gb.espacenet.com/).
[6] Wacky Patent of the Month (http://colitz.com/site.wacky.htm).
[7] Library of Congress, Copyright Office (http://www.loc.gov/copyright/).
[8] A Library of Congress Catalog Card Number (http://www.lcweb2.loc.gov/pcn/).
[9] International Standard Book Numbering (ISBN) (http://www.bowker.com/standards/).
[10] International Standard Serial Numbering (ISSN) (http://www.loc.gov/issn).

Additional Resources:

[11] J.L. Bryant, *Protecting Your Ideas: the Inventor's Guide to Patents*. San Diego. CA: Academic Press, 1999.
[12] S. Elias, *Patent, Copyright & Trademark*. Berkeley, CA: Nolo Press, 4th ed., 200 .
[13] F. Grissom and D. Pressman, *The Inventor's Notebook*. Berkeley, CA: Nolo Press. 2nd ed., 1996.
[14] D. Hitchcock, *Patent Searching Made Easy: How to Do Patent Searches on the Internet and in the Library*. Berkeley, CA: Nolo Press, 2nd ed., 2001.
[15] M. Lechter, Ed., *Successful Patents and Patenting for Engineers and Scientists*. New York: IEEE Press, 1995.
[16] American Bar Association, Section of Intellectual Property Law, Committee on Public Information, *Intellectual Property: A Guide for Engineers*. New York: ASME Press, 2001.
[17] General Information Concerning Patents. Washington, DC: United States Patent and Trademark Office, 2001, in (http://www.uspto.gov/web/offices/pac/doc/general/index.html).
[18] Guide to Filing a Utility Patent Application. Washington, DC: United States Patent and Trademark Office, 2001, in (http://www.uspto.gov/web/offices/pac/utility/utility.html).

Introduction

We start with the following question: ***What is an intellectual property?*** Intellectual property (IP) is a "product of mind" such as an idea, invention or a manufactured product. Or is an aggregate of rights resulting from creative efforts of the mind.

Intellectual property ***categories*** consist of patents, copyrights, plant protection (which include plant patents, plant variety certificates, utility plants, as well as plant breeder's rights overseas), trademarks, trade secrets and confidentiality (or secrecy) agreements, know-how/show-how. These are also various forms of legally protecting the intellectual property. The origin of the legally protecting these rights is in the United States Constitution that is reported next.

Under Article I, Section 8, Clause 8 of the United States Constitution: "The Congress shall have the power ... To promote the Progress of Science and the useful Arts, by securing for limited Times to Authors and Inventors the exclusive Right to their respective Writings and Discoveries." This is the basis for copyrights and patents laws. However, one must realize that the laws that are stemming from the above are not self-enforced, but rather one must protect his/her invention vigilantly. In other words, each individual in the United States must seek compensations when the underlying right have been infringed by others.

In this lecture we concentrate on most categories of intellectual properties other than patents. Subsequently, in Lecture Twenty Two we review issues pertaining to patents. What is common in these two lectures, however, is the method of bookkeeping that is described in Lecture Four. Just a reminder that that Laboratory Notebook, in whatever shape and form, must be maintained and archived regularly as described in Lecture Four. Here, the current and evolving thought and work performed must be recorded in an orderly manner that is ***easily verifiable***. The examiner for various IP applications inspects this Laboratory Notebook, and if it is found easily verifiable, then your chances of approval will increase substantially. The record must establish the priority of inventorship by documenting "when," "what," and "why," of the work performed, in a "diary" and ***not*** in a "memoir" style. Be particularly keen in providing some measure of proof against allegations of any impropriety and conflict of interest.

The forthcoming definitions below closely follow Ms. Elizabeth L. R. Donley's lecture.

I. Issues Pertaining to Copyrights

- We open this section by presenting the main ***purpose of copyright protection***.

    - Copyright encourages the creative efforts of authors, artists, and others by securing the exclusive right to reproduce works and derive income from them.
    - Copyright is created automatically once an original effort has been started and some aspect of that has been fixed in a tangible medium.
    - Copyright can also be registered for additional protection and notice to potential infringers.

- Next, the main ***Rights of Copyright Ownership.***

    - Reproduce, distribute and publicly display the work in various publishing media.
    - Prepare derivative works based on the original copyrighted work.
    - Publicly perform the work.

- ***What is Copyrightable?***

    - ***Tangible*** expressions are protected, but facts, ideas, concepts, principles, or discoveries are not.
    - Independent creation of a similar work is permitted to be copyrighted.

- **Copyrightable Subject Matter.**

  - Literary works.
  - Musical works.
  - Pictorial, Graphics and Sculptural works.
  - Architectural works.
  - Motion pictures and other audiovisual works.
  - Recently, Software and computer programs, as well as web pages.

- **Not- protected by Copyright.**

  - Ideas.
  - Titles.
  - Names.
  - Short phrases.
  - Works in public domain.
  - Mere facts.
  - Logos and slogans.
  - Blank forms for collecting information.
  - URL's.

In the United States ***Copyright Registration*** is necessary for recovering actual damages suffered from an unauthorized use of a work. The process is inexpensive and straightforward. For additional information please refer to [7]. In Appendix A, a set of samples forms for various copyright applications is shown for your information. These forms may change, of course, but the general requirements remain substantially the same. Therefore those interested to publish their report or any other work can see what they need.

- **Copyright Authorship**

  - Who is the "author?"
  - Is the work made-for-hire?
  - Is it a joint work?
  - Is it a compilation?
  - Is it a collective work?

- **Who Has Copyright Ownership?**

  - Original contributors.
  - Owners of pre-existing copyrighted material.
  - Other third parties such as employers (with work for hire clauses in employment agreements) or publishers.

- **Term of Copyright Protection**

  - Life of the author plus 70 years.
  - Work-for-hire – earlier of 75 years from publication or 100 years from creation.

- *Copyright Protection Weaknesses*

  - Does not prevent independent creation.
  - Does not protect purely functional aspects of the work.
  - Protection not available for ideas, procedures, processes, systems, methods of operation, concepts, principles, or discoveries.

- *The Fair Use Doctrine*

  - The "fair use" of a copyrighted work, for purposes such as criticism, comment, news reporting, teaching (including multiple copies for classroom use), scholarship, parody, or research, is not an infringement of copyright.
  - In other words, fair use is an affirmative defense to copyright infringement.

- *When is a use a "fair" use?* The courts use a four-part test:

  1. The purpose and character of the use, including whether such use is of a commercial nature or is for non-profit educational purposes.
  2. The nature of the copyrighted work.
  3. The amount and substantiality of the portion used in relation to copyrighted work as a whole.
  4. The effect of the use upon the potential market for or value of the copyrighted work.

- *Fair Use Factors*

  - The first factor asks primarily whether the work is of a commercial nature or whether it is intended for utilization in a non-profit educational setting.
  - The second factor looks to see whether the allegedly infringing work was created for the purpose of criticism, comment, news reporting, teaching, scholarship or research or for some other purpose such as commercial exploitation.
  - The third factor looks at how much of the work did someone uses or copy and whether that was the "heart" or "creative essence" of the work.
  - The forth and final factor addresses the effect of the use of the allegedly infringing work upon the "dollars and cents" aspect of the original work.
  - Fair use is almost always a short excerpt and almost always attributed. It should not harm the actual or potential commercial value of the work in the sense that people no longer need to buy it.

II. Issues Pertaining to Trade Secret

The formalities of trade secret are created by the individual State's status. Trade secret holders must take reasonable steps to protect their information. Never disclose a trade secret information to any outside party without first obtaining a signed *confidentiality disclosure agreement* (CDA). Trade secret protection and maintenance necessitates careful attention to employment agreements such as CDA's or Non-Compete Agreements or both.

- *Trade Secrets*

  - Information that is not generally known to competitors which provides a business advantage.
  - Revolves around secrecy.
  - Only protects against misappropriation.

III. Issues Pertaining to Know-how and Show-how

This is not the same as Trade Secret but is closely related or aligned to that. For instance, it can be very viable alternative even if not a secret when the cost of producing an item is greater than buying it from someone that knows how to make it and is willing to show how.

IV. Issues Pertaining to Trademark

Trademark Protection functions to increase distributional efficiency by making products easy to locate without any confusion. Trademark identifies the source of goods and services for customers. It also prevents competitors from using marks that may cause a likelihood of confusion.

- *Trademark Formalities*

    - Protection acquired through use in commerce, but one can file an intent to use application to mark place before product is even created.
    - Protections are available at both state and federal levels.
    - Common law trademark protection may be available in the absence of registration.

- *Term of Trademark Protection*

    - Generally, as long as the mark is used in commerce.
    - Must meet certain renewal requirements for registered trademarks.
    - Can renew intent to use application for three periods of six months before have to use in commerce.

- *Trademark Protection Weaknesses*

    - Does not in and of itself prevent misappropriation of the underlying product.
    - Must maintain some control over the quality of good or service.

Final Thought

Again, the following thought closely follows Ms. Elizabeth L. R. Donley's lecture.

- *Advice for Selecting IP Protection*

    - Since the costs of obtaining and enforcing IP's, for instance patents (to be presented in the next lecture note), are high, explore the availability and viability of other IP protection methods before proceeding further.
    - Have a patent search conducted by competent patent counsel *prior* to filing an application.
    - Does a patent *per se*, add value to the technology or is the maximum return likely to be achieved with copyright, trade secret protection or other IP protection?

Closure

- *What is a License Agreement?*

    - Basically an agreement by owners of IP not to sue others for infringing those IP rights.
    - Licenses are contracts in which owners of IP rights permit others to exercise all or some part of those rights under specific conditions.

- *Key Licensing Clauses* (To be prepared for an application.)

    - Grant Clauses.
        - License.
        - Grant Forward and Grant Backward.
    - License fee, royalty and other payment clauses.
    - Definition of licensed IP, field and territory.
    - Warranty and Indemnity Clause(s).
    - Termination Clause.

- *How to Capitalize on Technology*

    - Licensing an invention basically "rents" the use of an IP to a company.
    - Assignment of an IP in exchange for consideration.
    - Start-up company model capitalizes on added value.
    - Collaboration or joint ventures with industry.

Finally, we remind you to review the chart at the end of the next lecture note for an overall comparison of terms and possible protection tools for these intellectual properties.

Essential thoughts in this lecture

Issues.	Applicability to your project, if any.
Intellectual properties.	Is anyone of the above applicable to your project?
Do you want to add anything else?	Please elaborate.

Appendix A:

Attached is a set of sample forms for various copyright applications [7] to [10]. There are a host of other application forms for other types of Art which are not included here. This compilation is the courtesy of Ms. Nancy Spitzer of the K.F. Wendt Library at the University of Wisconsin-Madison.

IN ANSWER TO YOUR QUERY

FL 109

COPYRIGHT REGISTRATION OF BOOKS, MANUSCRIPTS, AND SPEECHES

LIBRARY OF CONGRESS

A published or unpublished book or manuscript may be submitted for registration in the Copyright Office. Form TX should be used to apply for copyright registration for textual works, with or without illustrations. Form TX is appropriate for registration of nondramatic literary works including: fiction, nonfiction, poetry, contributions to collective works, compilations, directories, catalogs, dissertations, theses, reports, speeches, bound or looseleaf volumes, pamphlets, brochures, and single pages containing text.

COPYRIGHT OFFICE

There is no specific requirement as to the printing, binding, format, or paper size and quality of unpublished manuscript material. Typewritten, photocopied, and legibly handwritten manuscripts, preferably in ink, are all acceptable for deposit. However, since deposit material represents the entire copyrightable content of a work submitted for registration, copies deposited in a format which would facilitate handling and long-term storage (e.g., stapled, bound, clipped, etc.) would be greatly appreciated by the Copyright Office.

101 Independence Avenue, S.E.

To register a book or manuscript, send the following three elements **in the same envelope or package** to the Library of Congress, Copyright Office, Register of Copyrights, 101 Independence Avenue, S.E., Washington, D.C. 20559-6000:

1. A completed application Form TX;
2. A nonrefundable filing fee of $30*
3. A nonreturnable deposit of the work. The deposit requirements depend on whether the work has been published at the time of registration:

Washington, D.C. 20559-6000

- If the work is unpublished, one complete copy or phonorecord.
- If the work was first published in the United States on or after January 1, 1978, two complete copies or phonorecords of the best edition.
- If the work was first published in the United States before January 1, 1978, two complete copies or phonorecords as first published.
- If the work was first published outside the United States, one complete copy or phonorecord of the work as first published.
- If the work is a contribution to a collective work, and published after January 1, 1978, one complete copy or phonorecord of the best edition of the collective work or a photocopy of the contribution itself as it was published in the collective work.

Copyright protects an author's expression in literary, artistic, or musical form. Copyright protection does not extend to any idea, system, method, device, name, or title.

Sincerely yours,

Register of Copyrights

---

* Fees are effective through June 30, 2002. After that date, check the Copyright Office Website at www.loc.gov/copyright or call (202) 707-3000 for current fee information.

How Long Does Copyright Registration Take?

A copyright registration is effective on the date of receipt in the Copyright Office of all required elements in acceptable form, regardless of the length of time it takes to process the application and mail the certificate of registration. The length of time required by the Copyright Office to process an application varies from time to time, depending on the amount of material received. Remember that it takes a number of days for mail to reach the Copyright Office and for the certificate of registration to reach the recipient after being mailed from the Copyright Office.

You will receive no acknowledgement that your application for copyright registration has been received (the Office receives more than 500,000 applications annually), but you may expect:
- A letter or telephone call from a Copyright Office staff member if further information is needed; and
- A certificate of registration to indicate the work has been registered, or if the application cannot be accepted, a letter explaining why it has been rejected.

You might not receive either of these until approximately 8 months after submission.

If you want to know when the Copyright Office received your material, send it via registered or certified mail and request a return receipt.

For further information, write:
Library of Congress
Copyright Office
Information Section, LM-401
101 Independence Ave., S.E.
Washington, D.C. 20559-6000

If you need additional application forms for copyright registration, call (202) 707-9100 at any time. Leave your request as a recorded message on the Copyright Office Forms and Publications Hotline in Washington, D.C. Please specify the kind and number of forms you need. If you have general information questions and wish to talk to an information specialist, call (202) 707-3000, TTY (202) 707-6737.

You may also photocopy blank application forms; **however,** photocopied forms submitted to the Copyright Office must be clear, legible, on a good grade of 8½-inch by 11-inch white paper suitable for automatic feeding through a scanner/photocopier. The forms should be printed, preferably in black ink, head-to-head (so that when you turn the sheet over, the top of page 2 is directly behind the top of page 1). **Forms not meeting these requirements will be returned to the originator.**

All U.S. Copyright Office application forms are available from the Copyright Office Website at **www.loc.gov/copyright**. They may be downloaded and printed for use in registering a claim to copyright or for use in renewing a claim to copyright.

You must have Adobe Acrobat Reader installed on your computer to view and print the forms. The free Adobe® Acrobat® Reader may be downloaded from Adobe Systems Incorporated through links from the same Internet site at which the forms are available.

Print forms head to head (top of page 2 is directly behind the top of page 1) on a single piece of good quality, 8½-inch by 11-inch white paper. To achieve the best quality copies of the application forms, use a laser printer.

Frequently requested Copyright Office circulars, announcements, and recently proposed as well as final regulations are also available from the Copyright Office Website.

Copyright Office circulars and announcements are available via fax. Call **(202) 707-2600** from any touchtone telephone. Key in your fax number at the prompt and the document number of the item(s) you want to receive by fax. The item(s) will be transmitted to your fax machine. If you do not know the document number of the item(s) you want, you may request that a menu be faxed to you. You may order up to three items at a time. Note that copyright application forms are *not* available by fax.

Library of Congress • Copyright Office • 101 Independence Avenue, S.E. • Washington, D.C. 20559-6000
http://www.loc.gov/copyright

Application Form TX

Detach and read these instructions before completing this form.
Make sure all applicable spaces have been filled in before you return this form.

BASIC INFORMATION

When to Use This Form: Use Form TX for registration of published or unpublished nondramatic literary works, excluding periodicals or serial issues. This class includes a wide variety of works: fiction, nonfiction, poetry, textbooks, reference works, directories, catalogs, advertising copy, compilations of information, and computer programs. For periodicals and serials, use Form SE.

Deposit to Accompany Application: An application for copyright registration must be accompanied by a deposit consisting of copies or phonorecords representing the entire work for which registration is to be made. The following are the general deposit requirements as set forth in the statute:
  Unpublished Work: Deposit one complete copy (or phonorecord)
  Published Work: Deposit two complete copies (or one phonorecord) of the best edition.
  Work First Published Outside the United States: Deposit one complete copy (or phonorecord) of the first foreign edition.
  Contribution to a Collective Work: Deposit one complete copy (or phonorecord) of the best edition of the collective work.
  The Copyright Notice: Before March 1, 1989, the use of copyright notice was mandatory on all published works, and any work first published before that date should have carried a notice. For works first published on and after March 1, 1989, use of the copyright notice is optional. For more information about copyright notice, see Circular 3, "Copyright Notices."

For Further Information: To speak to an information specialist, call (202) 707-3000 (TTY: (202) 707-6737). Recorded information is available 24 hours a day. Order forms and other publications from the address in space 9 or call the Forms and Publications Hotline at (202) 707-9100. Most circulars (but not forms) are available via fax. Call (202) 707-2600 from a touchtone phone. Access and download circulars, forms, and other information from the Copyright Office Website at www.loc.gov/copyright.

PRIVACY ACT ADVISORY STATEMENT Required by the Privacy Act of 1974 (P.L. 93-579)
  The authority for requesting this information is title 17, U.S.C., secs. 409 and 410. Furnishing the requested information is voluntary. But if the information is not furnished, it may be necessary to delay or refuse registration and you may not be entitled to certain relief, remedies, and benefits provided in chapters 4 and 5 of title 17, U.S.C.
  The principal uses of the requested information are the establishment and maintenance of a public record and the examination of the application for compliance with the registration requirements of the copyright code.
  Other routine uses include public inspection and copying, preparation of public indexes, preparation of public catalogs of copyright registrations, and preparation of search reports upon request.
  NOTE: No other advisory statement will be given in connection with this application. Please keep this statement and refer to it if we communicate with you regarding this application.

LINE-BY-LINE INSTRUCTIONS

Please type or print using black ink. The form is used to produce the certificate.

1 SPACE 1: Title

Title of This Work: Every work submitted for copyright registration must be given a title to identify that particular work. If the copies or phonorecords of the work bear a title or an identifying phrase that could serve as a title, transcribe that wording *completely* and *exactly* on the application. Indexing of the registration and future identification of the work will depend on the information you give here.

Previous or Alternative Titles: Complete this space if there are any additional titles for the work under which someone searching for the registration might be likely to look or under which a document pertaining to the work might be recorded.

Publication as a Contribution: If the work being registered is a contribution to a periodical, serial, or collection, give the title of the contribution in the "Title of This Work" space. Then, in the line headed "Publication as a Contribution," give information about the collective work in which the contribution appeared.

2 SPACE 2: Author(s)

General Instructions: After reading these instructions, decide who are the "authors" of this work for copyright purposes. Then, unless the work is a "collective work," give the requested information about every "author" who contributed any appreciable amount of copyrightable matter to this version of the work. If you need further space, request Continuation Sheets. In the case of a collective work, such as an anthology, collection of essays, or encyclopedia, give information about the author of the collective work as a whole.

Name of Author: The fullest form of the author's name should be given. Unless the work was "made for hire," the individual who actually created the work is its "author." In the case of a work made for hire, the statute provides that "the employer or other person for whom the work was prepared is considered the author."

What is a "Work Made for Hire"? A "work made for hire" is defined as (1) "a work prepared by an employee within the scope of his or her employment"; or (2) "a work specially ordered or commissioned for use as a contribution to a collective work, as a part of a motion picture or other audiovisual work, as a translation, as a supplementary work, as a compilation, as an instructional text, as a test, as answer material for a test, or as an atlas, if the parties expressly agree in a written instrument signed by them that the works shall be considered a work made for hire." If you have checked "Yes" to indicate that the work was "made for hire," you must give the full legal name of the employer (or other person for whom the work was prepared). You may also include the name of the employee along with the name of the employer (for example: "Elster Publishing Co., employer for hire of John Ferguson").

"Anonymous" or "Pseudonymous" Work: An author's contribution to a work is "anonymous" if that author is not identified on the copies or phonorecords of the work. An author's contribution to a work is "pseudonymous" if that author is identified on the copies or phonorecords under a fictitious name. If the work is "anonymous" you may: (1) leave the line blank; or (2) state "anonymous" on the line; or (3) reveal the author's identity. If the work is "pseudonymous" you may: (1) leave the line blank; or (2) give the pseudonym and identify it as such (for example: "Huntley Haverstock, pseudonym"); or (3) reveal the author's name, making clear which is the real name and which is the pseudonym (for example, "Judith Barton, whose pseudonym is Madeline Elster"). However, the citizenship or domicile of the author **must** be given in all cases.

Dates of Birth and Death: If the author is dead, the statute requires that the year of death be included in the application unless the work is anonymous or pseudonymous. The author's birth date is optional but is useful as a form of identification. Leave this space blank if the author's contribution was a "work made for hire."

Author's Nationality or Domicile: Give the country of which the author is a citizen or the country in which the author is domiciled. Nationality or domicile **must** be given in all cases.

Nature of Authorship: After the words "Nature of Authorship," give a brief general statement of the nature of this particular author's contribution to the work. Examples: "Entire text"; "Coauthor of entire text"; "Computer program"; "Editorial revisions"; "Compilation and English translation"; "New text."

SPACE 3: Creation and Publication

General Instructions: Do not confuse "creation" with "publication." Every application for copyright registration must state "the year in which creation of the work was completed." Give the date and nation of first publication only if the work has been published.

Creation: Under the statute, a work is "created" when it is fixed in a copy or phonorecord for the first time. Where a work has been prepared over a period of time, the part of the work existing in fixed form on a particular date constitutes the created work on that date. The date you give here should be the year in which the author completed the particular version for which registration is now being sought, even if other versions exist or if further changes or additions are planned.

Publication: The statute defines "publication" as "the distribution of copies or phonorecords of a work to the public by sale or other transfer of ownership, or by rental, lease, or lending." A work is also "published" if there has been an "offering to distribute copies or phonorecords to a group of persons for purposes of further distribution, public performance, or public display." Give the full date (month, day, year) when, and the country where, publication first occurred. If first publication took place simultaneously in the United States and other countries, it is sufficient to state "U.S.A."

SPACE 4: Claimant(s)

Name(s) and Address(es) of Copyright Claimant(s): Give the name(s) and address(es) of the copyright claimant(s) in this work even if the claimant is the same as the author. Copyright in a work belongs initially to the author of the work (including, in the case of a work made for hire, the employer or other person for whom the work was prepared). The copyright claimant is either the author of the work or a person or organization to whom the copyright initially belonging to the author has been transferred.

Transfer: The statute provides that, if the copyright claimant is not the author, the application for registration must contain "a brief statement of how the claimant obtained ownership of the copyright." If any copyright claimant named in space 4 is not an author named in space 2, give a brief statement explaining how the claimant(s) obtained ownership of the copyright. Examples: "By written contract"; "Transfer of all rights by author"; "Assignment"; "By will." Do not attach transfer documents or other attachments or riders.

SPACE 5: Previous Registration

General Instructions: The questions in space 5 are intended to show whether an earlier registration has been made for this work and, if so, whether there is any basis for a new registration. As a general rule, only one basic copyright registration can be made for the same version of a particular work.

Same Version: If this version is substantially the same as the work covered by a previous registration, a second registration is not generally possible unless: (1) the work has been registered in unpublished form and a second registration is now being sought to cover this first published edition; or (2) someone other than the author is identified as copyright claimant in the earlier registration, and the author is now seeking registration in his or her own name. If either of these two exceptions applies, check the appropriate box and give the earlier registration number and date. Otherwise, do not submit Form TX. Instead, write the Copyright Office for information about supplementary registration or recordation of transfers of copyright ownership.

Changed Version: If the work has been changed and you are now seeking registration to cover the additions or revisions, check the last box in space 5, give the earlier registration number and date, and complete both parts of space 6 in accordance with the instructions below.

Previous Registration Number and Date: If more than one previous registration has been made for the work, give the number and date of the latest registration.

SPACE 6: Derivative Work or Compilation

General Instructions: Complete space 6 if this work is a "changed version," "compilation," or "derivative work" and if it incorporates one or more earlier works that have already been published or registered for copyright or that have fallen into the public domain. A "compilation" is defined as "a work formed by the collection and assembling of preexisting materials or of data that are selected, coordinated, or arranged in such a way that the resulting work as a whole constitutes an original work of authorship." A "derivative work" is "a work based on one or more preexisting works." Examples of derivative works include translations, fictionalizations, abridgments, condensations, or "any other form in which a work may be recast, transformed, or adapted." Derivative works also include works "consisting of editorial revisions, annotations, or other modifications" if these changes, as a whole, represent an original work of authorship.

Preexisting Material (space 6a): For derivative works, complete this space **and** space 6b. In space 6a identify the preexisting work that has been recast, transformed, or adapted. The preexisting work may be material that has been previously published, previously registered, or that is in the public domain. An example of preexisting material might be: "Russian version of Goncharov's 'Oblomov.'"

Material Added to This Work (space 6b): Give a brief, general statement of the new material covered by the copyright claim for which registration is sought. **Derivative work** examples include: "Foreword, editing, critical annotations"; "Translation"; "Chapters 11-17." If the work is a **compilation**, describe both the compilation itself and the material that has been compiled. Example: "Compilation of certain 1917 Speeches by Woodrow Wilson." A work may be both a derivative work and compilation, in which case a sample statement might be: "Compilation and additional new material."

SPACE 7, 8, 9: Fee, Correspondence, Certification, Return Address

Deposit Account: If you maintain a Deposit Account in the Copyright Office, identify it in space 7a. Otherwise leave the space blank and send the fee of $30 (effective through June 30, 2002) with your application and deposit.

Correspondence (space 7b): This space should contain the name, address, area code, telephone number, fax number, and email address (if available) of the person to be consulted if correspondence about this application becomes necessary.

Certification (space 8): The application cannot be accepted unless it bears the date and the **handwritten signature** of the author or other copyright claimant, or of the owner of exclusive right(s), or of the duly authorized agent of author, claimant, or owner of exclusive right(s).

Address for Return of Certificate (space 9): The address box must be completed legibly since the certificate will be returned in a window envelope.

Part Two: Lecture Twenty One – Intellectual Property Issues Page 133

FEE CHANGES
Fees are effective through June 30, 2002. After that date, check the Copyright Office Website at www.loc.gov/copyright or call (202) 707-3000 for current fee information.

FORM TX
For a Nondramatic Literary Work
UNITED STATES COPYRIGHT OFFICE
REGISTRATION NUMBER

_____

TX   TXU
EFFECTIVE DATE OF REGISTRATION

_____
Month   Day   Year

DO NOT WRITE ABOVE THIS LINE. IF YOU NEED MORE SPACE, USE A SEPARATE CONTINUATION SHEET.

1 TITLE OF THIS WORK ▼

PREVIOUS OR ALTERNATIVE TITLES ▼

PUBLICATION AS A CONTRIBUTION If this work was published as a contribution to a periodical, serial, or collection, give information about the collective work in which the contribution appeared.   Title of Collective Work ▼

If published in a periodical or serial give: Volume ▼   Number ▼   Issue Date ▼   On Pages ▼

2 a NAME OF AUTHOR ▼   DATES OF BIRTH AND DEATH
Year Born ▼   Year Died ▼

Was this contribution to the work a "work made for hire"?
☐ Yes
☐ No

AUTHOR'S NATIONALITY OR DOMICILE
Name of Country
OR { Citizen of ▶ _____
     Domiciled in ▶ _____

WAS THIS AUTHOR'S CONTRIBUTION TO THE WORK
Anonymous?   ☐ Yes ☐ No
Pseudonymous? ☐ Yes ☐ No
If the answer to either of these questions is "Yes," see detailed instructions.

NATURE OF AUTHORSHIP Briefly describe nature of material created by this author in which copyright is claimed. ▼

NOTE
Under the law, the "author" of a "work made for hire" is generally the employer, not the employee (see instructions). For any part of this work that was "made for hire" check "Yes" in the space provided, give the employer (or other person for whom the work was prepared) as "Author" of that part, and leave the space for dates of birth and death blank.

b NAME OF AUTHOR ▼   DATES OF BIRTH AND DEATH
Year Born ▼   Year Died ▼

Was this contribution to the work a "work made for hire"?
☐ Yes
☐ No

AUTHOR'S NATIONALITY OR DOMICILE
Name of Country
OR { Citizen of ▶ _____
     Domiciled in ▶ _____

WAS THIS AUTHOR'S CONTRIBUTION TO THE WORK
Anonymous?   ☐ Yes ☐ No
Pseudonymous? ☐ Yes ☐ No
If the answer to either of these questions is "Yes," see detailed instructions.

NATURE OF AUTHORSHIP Briefly describe nature of material created by this author in which copyright is claimed. ▼

c NAME OF AUTHOR ▼   DATES OF BIRTH AND DEATH
Year Born ▼   Year Died ▼

Was this contribution to the work a "work made for hire"?
☐ Yes
☐ No

AUTHOR'S NATIONALITY OR DOMICILE
Name of Country
OR { Citizen of ▶ _____
     Domiciled in ▶ _____

WAS THIS AUTHOR'S CONTRIBUTION TO THE WORK
Anonymous?   ☐ Yes ☐ No
Pseudonymous? ☐ Yes ☐ No
If the answer to either of these questions is "Yes," see detailed instructions.

NATURE OF AUTHORSHIP Briefly describe nature of material created by this author in which copyright is claimed. ▼

3 a YEAR IN WHICH CREATION OF THIS WORK WAS COMPLETED   This information must be given ◀ Year in all cases.

b DATE AND NATION OF FIRST PUBLICATION OF THIS PARTICULAR WORK
Complete this information ONLY if this work has been published.   Month ▶ _____ Day ▶ _____ Year ▶ _____
◀ Nation

4 COPYRIGHT CLAIMANT(S) Name and address must be given even if the claimant is the same as the author given in space 2. ▼

See instructions before completing this space.

TRANSFER If the claimant(s) named here in space 4 is (are) different from the author(s) named in space 2, give a brief statement of how the claimant(s) obtained ownership of the copyright. ▼

APPLICATION RECEIVED
_____
ONE DEPOSIT RECEIVED
_____
TWO DEPOSITS RECEIVED
_____
FUNDS RECEIVED
_____
DO NOT WRITE HERE OFFICE USE ONLY

MORE ON BACK ▶   • Complete all applicable spaces (numbers 5-9) on the reverse side of this page.
• See detailed instructions.   • Sign the form at line 8.

DO NOT WRITE HERE
Page 1 of _____ pages

EXAMINED BY	FORM TX
CHECKED BY	
☐ CORRESPONDENCE Yes	FOR COPYRIGHT OFFICE USE ONLY

DO NOT WRITE ABOVE THIS LINE. IF YOU NEED MORE SPACE, USE A SEPARATE CONTINUATION SHEET.

PREVIOUS REGISTRATION Has registration for this work, or for an earlier version of this work, already been made in the Copyright Office?
☐ Yes ☐ No If your answer is "Yes," why is another registration being sought? (Check appropriate box.) ▼
a. ☐ This is the first published edition of a work previously registered in unpublished form.
b. ☐ This is the first application submitted by this author as copyright claimant.
c. ☐ This is a changed version of the work, as shown by space 6 on this application.
If your answer is "Yes," give: **Previous Registration Number** ▶   **Year of Registration** ▶

5

DERIVATIVE WORK OR COMPILATION
Preexisting Material Identify any preexisting work or works that this work is based on or incorporates. ▼

a

6

Material Added to This Work Give a brief, general statement of the material that has been added to this work and in which copyright is claimed. ▼

b

See instructions before completing this space.

DEPOSIT ACCOUNT If the registration fee is to be charged to a Deposit Account established in the Copyright Office, give name and number of Account.
Name ▼   **Account Number** ▼

a

7

CORRESPONDENCE Give name and address to which correspondence about this application should be sent.   Name/Address/Apt/City/State/ZIP ▼

b

Area code and daytime telephone number ▶   Fax number ▶
Email ▶

CERTIFICATION* I, the undersigned, hereby certify that I am the
Check only one ▶ { ☐ author
☐ other copyright claimant
☐ owner of exclusive right(s)
☐ authorized agent of _____
of the work identified in this application and that the statements made   Name of author or other copyright claimant, or owner of exclusive right(s) ▲
by me in this application are correct to the best of my knowledge.

8

Typed or printed name and date ▼ If this application gives a date of publication in space 3, do not sign and submit it before that date.
_____ Date ▶ _____

Handwritten signature (X) ▼
X _____

| Certificate will be mailed in window envelope to this address: | Name ▼

Number/Street/Apt ▼

City/State/ZIP ▼ | **YOU MUST:**
• Complete all necessary spaces
• Sign your application in space 8
SEND ALL 3 ELEMENTS IN THE SAME PACKAGE:
1. Application form
2. Nonrefundable filing fee in check or money order payable to *Register of Copyrights*
3. Deposit material
MAIL TO:
Library of Congress
Copyright Office
101 Independence Avenue, S.E.
Washington, D.C. 20559-6000 | As of July 1, 1999, the filing fee for Form TX is $30. |

9

*17 U.S.C. § 506(e): Any person who knowingly makes a false representation of a material fact in the application for copyright registration provided for by section 409, or in any written statement filed in connection with the application, shall be fined not more than $2,500.
June 1999—200,000   ♲ PRINTED ON RECYCLED PAPER   ☆U.S. GOVERNMENT PRINTING OFFICE: 1999-454-879/49
WEB REV: June 1999

ISBN U.S. Agency and Its Functions

The U.S. ISBN Agency is responsible for the assignment of the ISBN Publisher Prefix to those publishers with a residence or office in the U.S. and are publishing their titles within the U.S.

It cannot assign ISBNs to foreign publishers (see ISBN Group Agencies for foreign local agency addresses).

The U.S. agency address is:
> U.S. ISBN Agency
> 630 Central Avenue
> New Providence, NJ 07974
> Tel: 877-310-7333
> Fax: 908-665-2895
> isbn-san@bowker.com

The ISBN agency allocates and assigns publisher prefixes (which are unique to each publisher).

It supplies a list of Bookland EAN Bar Code Film Master Suppliers and hyphenation instructions and provides information on other related resources which publishers may require, including Cataloging in Publication program (CIP), Library of Congress Catalog Control Number (LCCN), International Standard Serial Number (ISSN), PUBNET, and the Book Industry Study Group (BISG).

The agency encourages and promotes the importance of the ISBN for proper bibliographic listing in Books in Print, Forthcoming Books and other directories.

The agency also maintains an up-to-date and comprehensive listing of all U.S. Publisher Prefix assignments for inclusion in the Publishers, Distributors & Wholesalers of the United States. This information is taken from R.R. Bowker's Publishers Authority Database.

HOME | ABOUT | FEEDBACK | HELP | RESOURCES | ISBN US | ISBN PR | ISBN INT'L | ISMN | DIGITAL ISBN/DEV

Bowker

Copyright © 2001 R.R. Bowker LLC. All rights reserved.

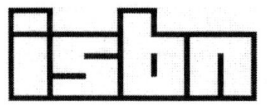

INTERNATIONAL STANDARD BOOK NUMBER--UNITED STATES AGENCY
International Standard Numbering System for the Information Industry
630 Central Avenue, New Providence, New Jersey 07974
TEL: 908-771-7755 FAX: 908-665-2895 Email: isbn-san@bowker.com
International Standard ISO 2108   R.R. Bowker LLC, a Cambridge Information Group company

APPLICATION FOR AN ISBN PUBLISHER PREFIX

FOR AGENCY USE ONLY

SYMBOL: _____

PREFIX: _____

PLEASE PRINT OR TYPE:

Company/Publisher Name: _____

Address: _____

Phone Number: _____ Do Not Publish:

Fax Number: _____ Do Not Publish:

Toll Free Number: _____

E-MAIL: _____ Do Not Publish:

Web Site: _____

Fax-on-Demand: _____ Toll Free Fax: _____

If P.O. Box Indicated, Local Street Address is Required:

Name of Rights & and Permissions Contact: _____

Title: _____ Phone Number: _____

Name of ISBN Coordinator/Contact: _____

Title: _____ Phone Number: _____

Division or Subsidiary of: _____

Imprints: _____

APPLICATION FOR AN ISBN PUBLISHER PREFIX

PAYMENT: <u>A NON-REFUNDABLE PROCESSING SERVICE CHARGE</u>
<u>PRIORITY PROCESSING SURCHARGE $75</u>

ISBN PREFIX BLOCK	REGULAR PROCESSING FEE	PRIORITY PROCESSING FEE
10 ISBNs	$225.00	$300.00
100 ISBNs	$800.00	$875.00
1,000 ISBNs	$1,200.00	$1,275.00
10,000 ISBNs	$3,000.00	-

Fee Waiver:
Applicants requesting a fee-waiver MUST provide a list of titles and formats along with 501(C3) and mission statement documents. Failure to provide this title list will delay Agency processing.

__ Check/Money Order enclosed.   Make payable to "R. R. Bowker."

__ Charge: ____ American Express   ____ Visa   ____ Master Card

Card Holder Name: _____

    Account #: _____   Expiration Date: _____

Total amount enclosed or charged: _____

Authorized signature: _____   Date: _____

* <u>Note</u>: **Credit Cards are the preferred form of payment**

<u>PUBLISHING INFORMATION</u>:

1. Indicate year you started publishing: _____

2. Indicate what type of products you produce (circle):

    Books   Videos   Spoken Words on Cassette/CD
         Software   Mixed Media

    Other - Please specify: _____

3. Book Subject Area (circle):

    o Children's
    o Law
    o Medical
    o Religious
    o Sci-Tech
    o Other - Please specify: _____

<u>DISTRIBUTION INFORMATION</u>:

1. Do you distribute for, or are you distributed by, any other company?
    Yes: _____   No: _____. If yes, please provide full company name

```
address and ISBN Publisher Prefix (if any):
```
_____
_____
_____

PROCESSING INFORMATION:

Your application for an ISBN Publisher Prefix will be processed ONLY if you include the following:

1. Completed application
2. Payment

* **Note: Credit Cards are the preferred form of payment**

The ISBN U.S. Agency will not provide an ISBN by telephone or fax. Processing time for an ISBN application is 10 business days (Saturdays and Sundays and holidays are not business days) from the date of our receipt of the completed form. This means that the application is inhouse for that length of time; ISBNs will be mailed to publishers after this processing period is completed (provided there are no problems with the application).

PRIORITY PROCESSING:

If you intend to ask the agency for a faster turn around time, a priority charge of $75. applies and must be added to the service charge fee. Priority service includes return, within 48 business hours of receipt, of your ISBN Publisher Prefix and ISBN log book (provided there are no problems with the application).

If you are requesting priority service and would like your ISBN log book e-mailed, please provide the e-mail address to where it should be sent: _____

Please Note: The priority service is either by e-mail or courier service, but NOT both.

WAIVING OF THE SERVICE CHARGE:

Your firm may apply for a waiver of the service charge if your firm has been granted a 501 (C3) charitable/philanthropic tax exemption status & your firm can supply a statement of your charitable/philanthropic mission. Your firm must supply documentation on BOTH to be eligible for a fee waiver. If you request a waiver of the service charge and require priority processing, a charge of $75.00 does apply.

```
              Return the application and payment to:

              ISBN U.S. Agency
              R.R. Bowker
              630 Central Avenue
              New Providence, NJ 07974
              isbn-san@bowker.com
```

LECTURE TWENTY TWO: *Intellectual Property and Patents*

Remark: This lecture note is the continuation of the preceding one and as such shares the special resources and additional references of that lecture note. In the following issues pertaining to intellectual property and preliminary steps to apply for patents are discussed. Again, the purpose of this and the previous lecture note is to clarify these matters and give further information on what some of these terminologies mean and how to seek a patent. Clearly, and as stated before, it is not our intention to serve as an attorney nor do we plan to provide a plethora of forms and documents in this regard. Special credit in preparation of this lecture note, is the same as the preceding note.

Introduction

In this lecture we concentrate on issues pertaining to patents. Again, what is common in these topics, however, is the method of bookkeeping that is described in Lecture Four. Just a reminder that this laboratory notebook in whatever shape and form must be maintained and archived regularly as described. Here, the current and evolving thoughts and work performed must be recorded in an orderly manner that is *easily verifiable*. The patent examiner wants it that way. The record must establish the priority of inventorship. Be particularly careful in providing proofs against allegations of any impropriety and conflict of interest.

I. Issues Pertaining to Patents

In the following we concentrate on issues pertaining to patents. The forthcoming definitions in subsequent sections of this note closely follow Ms. Deanna L. Dietrich and Ms. Nancy Spitzer's lecture, with some input from Dr. Steven Price's lecture.

What is a patent? A legal right that permits the owner to exclude others from making, using, and selling an invention. A patent is granted to individuals who invent *new* and *useful* inventions. Here, the keywords are *new* and *useful*. During a patent's *limited term*, its owner has the right to exclude others from making, using, selling, offering for sale or importing the patented invention. Exclusionary rights lasts for 20 years from the date of the patent application was filed (below). In return, the inventor must disclose his or her invention to the world in such detail that other people will be able to make and use it. When a patent is issued, someone else may also have part of that idea. In that case, a patent owner may be prohibited from using his/her own invention if an earlier patent covers a portion of that invention. In some instances, some other laws may prohibit practicing an invention, which in that case the inventor may need permission to use his/her own invention. Thus, the patent alone does not give the owner the rights to make, use, sell offer for sale or import an invention.

What is the life of a Patent? Effective June 6, 1995, the patent law changed so that all patents filed on or after that date will be in effect (called its "term") for 20 years from the date the patent application was *filed* with the United States Patent and Trademark Office. This has replaced the previous term of 17 years from the *issue date* of the patent.

What is Prior Art? General state of knowledge existing or publicly available at least one year prior to patent application date.

What are patentable subject matters? Generally, there are three major types of patents.

- **Utility Patents:** referring to processes, machines, manufactures, composition of matter.
- **Plant Patents:** referring to asexually reproduced plants.
- **Design Patents:** referring to ornamental designs.
- Recently, **software patents** are also being entertained as part of utility patents.

There are certainly items, which *cannot* be patented. A partial list is as follows.

- **An inoperative device.**
- **Inventions that are already known in the prior Art.**
- **Laws of nature.**
- **Mathematical formulas.**
- **Method of doing business.**
- **Printed material.**

II. Preliminary Steps Towards Securing Patents

Determining the patentability of an invention is certainly the first step in a long and arduous process of securing a patent. The patent examiner in the United States will look at an application and ask the following questions (From Patent Status, Title 35 U.S. Code, Sections 101, 102,103)

- *Is this the first person to invent?* In some other countries the examiner looks at the first person that applies. One cannot patent an invention in the U.S. if it is known or used by the public more than one year prior to the date of a patent application being filed. In foreign countries, one cannot obtain a foreign patent if there is any public disclosure before a patent application is filed. Public disclosure includes seminars, poster sessions, and articles.

- *Utility – Is this a useful subject matter?* "… Useful process, machine, manufacture, or composition of matter, or any new and useful improvement thereof…" Usefulness of the invention plays a central role in this process.

- *Novelty – Is this a novel approach?* Novel means no prior work of that type is out there. Cannot patent an invention if it is known or used by the public more than one year prior to the date of the patent application. Do not publish until your application is ready. In foreign countries it pays to be the first to apply, but do not publish about the work until all applications are ready. Unless one is seeking a *provisional* patent.

- *Non-obvious?* This means the invention must not be the next logical step in the area and cannot be an obvious extension of what is already known to a person having ordinary skill in the art."

Provisional Application: In order to secure an earlier filing date, effective June 8, 1995, a Provisional Patent Application can be filed. A Provisional Application is a simpler form of a patent application, enabling rapid securing of a filing date. If a provisional Application is filed, a complete application must be filed within one year of the provisional filing date.

Important dates to remember: The patent examiner looks into the proof, or lack of it, of the following dates.

(1) When is the earliest date of "conception?"
(2) When is the earliest date of "reduction to practice?"
(3) When is the first date of public use?
(4) When is the first publication date?
(5) When is the first date of offer for sale?
(6) When is the date for the first sale?

It is clear that one tool to establish the above is the "Laboratory Notebook," which must have been developed throughout this process.

- ***Publishing and Seminars***

  - In the U.S., a person has one year from the time of public disclosure of an invention to file a patent application.
  - However, this "grace period" is not recognized outside the U.S., and in most foreign countries one cannot obtain a foreign patent if public enabling occurs before filing a U.S. patent.
  - Therefore, it is recommended to begin the patent process before one discloses an idea.
  - Provisional patent applications are expected to be very helpful because if an applicant is about to make a public disclosure, the filing of a provisional application, which in theory can be done rapidly and with limited expense, will preserve an early filing date.

III. Secondary Steps - Searching

Searching is the second logical step to entertain before a formal application is submitted to the patent office. In this phase we study the following concerns by searching for prior Arts and patents through resources listed above, particularly [2].

- ***Has this invention already been patented?*** This is done through a novelty search [2]. This is difficult to perform since it seeks to prove a negative – an invention has not existed previously. When reviewing this site, and searching for patents, consider the following very carefully as the immediate benefits of this search.

  - ***Look at the sample patents to help prepare the application.***
    Get valuable background information on prior Art that you can use in your application.

  - ***Get valuable technical information from similar patents.***
    This helps you to understand the current and prior state of technology – That whether you have really invented a patentable project.

  - ***Help identify the novel features of your invention.***

  - ***Search patent applications for similar "pending patents."***
    Now, you search U.S. patent applications since March 2001. Applications are made public if the invention is also filed in a foreign country. The application may not be granted, but you can see what is out there.

Using the US patent classification, we search for patents by subject. First, determine patent classification and sub-classification numbers using the Manual of Classification. This is the most precise and comprehensive way to search the patent database looking for prior Art.

Then search patent data base by class number, and read relevant patents. This system consists of 300 main classes and 120,000 subclasses.

Each class represents an invention's function, structure, substance or field of use.

This is used by patent examiners to categorize every new application and to do a prior Art search.

It is detailed, hierarchical and fast changing (can expand or be reorganized over time).

Or, search patent data base by keyword (back to 1976) first, then locate relevant patents and appropriate class numbers. Continue the search by classification number to retrieve patents back to 1790.

- *Patent Prosecution*

    - Upon receipt of a favorable patentability opinion, together with a consideration of the inventions' commercial utility, the attorney may be instructed to draft a patent application covering the invention.
    - In some instances further work may be required of the inventors or others so that a proper application can be drafted.
    - After a period usually lasting several months and including a detailed review of the final draft, an application will be submitted to the U.S. Patent and Trademark Office.
    - Inventor has one year from filing date to decide on foreign filing.

IV. Final Steps Disclosure and Application

The process consists of the following steps, which must be completed by a competent counsel. However, the applicant must compile the necessary information listed below.

Step 1 – Patent Disclosure

- The patent journey begins when you believe you have made a discovery.
- Discoveries are summarized in an "Invention Disclosure Report."

Step 2 – The Invention Disclosure Report

This report asks for the following entries that must be prepared by the applicant, just as the final report for any project.
- A good descriptive title.
- A simple description of the invention and date it was made.
- A description of all the ways it could be used.
- Recent literature including talks that have been given and reports that have been prepared.
- Names of all inventors.
- Names of others to whom you have disclosed the invention.
- Funding sources (work-for-hire?).
- Signatures.

Step 3 – A Patent Application

This application asks for the following information (a sample application is attached).

- *Title*
- *Inventor(s)* – Can be more than one inventor or *"patentee."*
- *Assignee (Applicant)* – Company or person to whom the patentee has assigned the patent rights usually because he/she is an employee of that company.
- *Classification Numbers* – This is organized by the USPTO, currently into 300 main classes and 120000 subclasses. There is an international class system called the IPC, and foreign patent can be searched using those numbers.
- *Prior Art References* – Citations to other patents, journal articles or books as well as other patents that are closely related to this application.
- *Abstract* – Summary of the technical disclosures made in the patent.
- *Specification* – Consists of the description of the claims.
- *Claims* - Describe the structure of an invention in precise terms and in one sentence define the invention. The more precise, the better is received by the examiner. Describe the subject matter and from the legal description and boundaries of invention. This is used to set the basis for an infringement suit.
- *Drawings*

Step 4 – Plan of Action

Steps for obtaining a patent culminates in a package containing the following segments:

- Invention disclosure report.
- Patent application that is prepared by an attorney.
- Patent applications sent to U.S. patent and Trademark Office for "filing." Patent is now "pending."

Then

- Patent prosecuted, "allowed," is issued – possible grant is made to the inventor, publications can move forward.

Final Thought

The importance of a well-prepared and well-archived "Laboratory Notebook" in this entire process cannot be overemphasized. Having that notebook is of paramount importance in successfully obtaining a patent, or any IP license.

Closure

It is very important that we start with a patent search in [2], in order to get the "Big Picture" and see how this invention will add to the state of technology before spending additional time and money to seek a license. Also, one needs a competent counsel to finish the task. Thus, ask for an attorney with experience in this field. A summary of relationships among these IP's is given subsequently.

Essential thoughts in this lecture

Issues.	Applicability to your project, if any.
Patents.	Is your project patentable?
Do you want to add anything else?	Please elaborate.

Relationships among Trade Secrets, Patents, Trade Names, Trademarks and Copyrights

	Origin of Rights	Prerequisites to Protection	Scope of Protection	Life	Test for Infringement
Trade Secret	Investment of time and money.	Recognition of value and utility, access.	Confidential subject matter.	Life of confidentiality.	Derivation.
Utility patent	Grant by Federal Government, on application by inventor.	New, useful, and non-obvious subject matter.	Useful process, machine article of manufacture, or composition of matter.	17 years from date of grant. Since June 6, 1995, 20 years from the date filed.	Manufacture, use or for sale in U.S.A., of claimed invention, including use or sale of invention made outside U.S.A., by a process patented in the U.S.A.
Design Patent	Grant by Federal Government, on application by inventor.	New, original and ornamental subject matter.	Ornamental design for article of manufacture	14 years from date of grant.	Designs look alike to eye of an ordinary observer.
Copyright	Creation of "works of authorship".	Originality – registration before infringement or within three months of publication required for attorney fees and statutory remedies.	Works of authorship.	Variable – on the order 50 years or longer; life of author plus 50 years.	Copying.
Trade Name Trademark Service mark	Adoption and use, or intent to use coupled with registration.	Use to identify and distinguish business goods or services.	Words, names, symbols, or devices.	As long as property used as a mark.	Likelihood of confusion, mistake, or deception.

Appendix A:

Attached is a set of sample forms for patent application [2]. This compilation is the courtesy of Ms. Nancy Spitzer of the K.F. Wendt Library at the University of Wisconsin-Madison.

Part Two: Lecture Twenty Two – Intellectual Property Issues   *Page 145*

PTO/SB/01 (10-01)
Approved for use through 10/31/2002. OMB 0651-0032
U.S. Patent and Trademark Office; U.S. DEPARTMENT OF COMMERCE

Under the Paperwork Reduction Act of 1995, no persons are required to respond to a collection of information unless it contains a valid OMB control number.

DECLARATION FOR UTILITY OR DESIGN PATENT APPLICATION (37 CFR 1.63)

Attorney Docket Number	
First Named Inventor	
COMPLETE IF KNOWN	
Application Number	
Filing Date	
Art Unit	
Examiner Name	

☐ Declaration Submitted with Initial Filing   **OR**   ☐ Declaration Submitted after Initial Filing (surcharge (37 CFR 1.16 (e)) required)

As the below named inventor, I hereby declare that:

My residence, mailing address, and citizenship are as stated below next to my name.

I believe I am the original and first inventor of the subject matter which is claimed and for which a patent is sought on the invention entitled:

[   ]

(Title of the Invention)

the specification of which

☐ is attached hereto

OR

☐ was filed on (MM/DD/YYYY) [   ] as United States Application Number or PCT International

Application Number [   ] and was amended on (MM/DD/YYYY) [   ] (if applicable).

I hereby state that I have reviewed and understand the contents of the above identified specification, including the claims, as amended by any amendment specifically referred to above.

I acknowledge the duty to disclose information which is material to patentability as defined in 37 CFR 1.56, including for continuation-in-part applications, material information which became available between the filing date of the prior application and the national or PCT international filing date of the continuation-in-part application.

I hereby claim foreign priority benefits under 35 U.S.C. 119(a)-(d) or (f), or 365(b) of any foreign application(s) for patent, inventor's or plant breeder's rights certificate(s), or 365(a) of any PCT international application which designated at least one country other than the United States of America, listed below and have also identified below, by checking the box, any foreign application for patent, inventor's or plant breeder's rights certificate(s), or any PCT international application having a filing date before that of the application on which priority is claimed.

Prior Foreign Application Number(s)	Country	Foreign Filing Date (MM/DD/YYYY)	Priority Not Claimed	Certified Copy Attached? YES NO
			☐	☐ ☐
			☐	☐ ☐
			☐	☐ ☐
			☐	☐ ☐

☐ Additional foreign application numbers are listed on a supplemental priority data sheet PTO/SB/02B attached hereto:

[Page 1 of 2]

Burden Hour Statement: This form is estimated to take 21 minutes to complete. Time will vary depending upon the needs of the individual case. Any comments on the amount of time you are required to complete this form should be sent to the Chief Information Officer, U.S. Patent and Trademark Office, Washington, DC 20231. DO NOT SEND FEES OR COMPLETED FORMS TO THIS ADDRESS. SEND TO: Assistant Commissioner for Patents, Washington, DC 20231.

PTO/SB/01 (10-01)
Approved for use through 10/31/2002. OMB 0651-0032
U.S. Patent and Trademark Office; U.S. DEPARTMENT OF COMMERCE
Under the Paperwork Reduction Act of 1995, no persons are required to respond to a collection of information unless it contains a valid OMB control number.

DECLARATION — Utility or Design Patent Application

Direct all correspondence to: ☐ Customer Number or Bar Code Label [_____] OR ☐ Correspondence address below

Name

Address

City	State	ZIP

Country	Telephone	Fax

I hereby declare that all statements made herein of my own knowledge are true and that all statements made on information and belief are believed to be true; and further that these statements were made with the knowledge that willful false statements and the like so made are punishable by fine or imprisonment, or both, under 18 U.S.C. 1001 and that such willful false statements may jeopardize the validity of the application or any patent issued thereon.

NAME OF SOLE OR FIRST INVENTOR : ☐ A petition has been filed for this unsigned inventor

Given Name (first and middle [if any])	Family Name or Surname

Inventor's Signature		Date

Residence: City	State	Country	Citizenship

Mailing Address

City	State	ZIP	Country

NAME OF SECOND INVENTOR: ☐ A petition has been filed for this unsigned inventor

Given Name (first and middle [if any])	Family Name or Surname

Inventor's Signature		Date

Residence: City	State	Country	Citizenship

Mailing Address

City	State	ZIP	Country

☐ Additional inventors are being named on the ____ supplemental Additional Inventor(s) sheet(s) PTO/SB/02A attached hereto.

[Page 2 of 2]

PTO/SB/16 (10-01)
Approved for use through 10/31/2002. OMB 0651-0032
U.S. Patent and Trademark Office; U.S. DEPARTMENT OF COMMERCE
Under the Paperwork Reduction Act of 1995, no persons are required to respond to a collection of information unless it displays a valid OMB control number.

PROVISIONAL APPLICATION FOR PATENT COVER SHEET
This is a request for filing a PROVISIONAL APPLICATION FOR PATENT under 37 CFR 1.53(c).

Express Mail Label No.

INVENTOR(S)

Given Name (first and middle [if any])	Family Name or Surname	Residence (City and either State or Foreign Country)

☐ Additional inventors are being named on the _____ separately numbered sheets attached hereto

TITLE OF THE INVENTION (500 characters max)

CORRESPONDENCE ADDRESS
Direct all correspondence to:

☐ Customer Number — *Type Customer Number here* → *Place Customer Number Bar Code Label here*

OR

☐ Firm *or* Individual Name

Address	
Address	
City	State	ZIP
Country	Telephone	Fax

ENCLOSED APPLICATION PARTS *(check all that apply)*

☐ Specification   Number of Pages ____
☐ Drawing(s)   Number of Sheets ____
☐ Application Data Sheet. See 37 CFR 1.76
☐ CD(s), Number ____
☐ Other (specify) ____

METHOD OF PAYMENT OF FILING FEES FOR THIS PROVISIONAL APPLICATION FOR PATENT

☐ Applicant claims small entity status. See 37 CFR 1.27.
☐ A check or money order is enclosed to cover the filing fees
☐ The Commissioner is hereby authorized to charge filing fees or credit any overpayment to Deposit Account Number: ____
☐ Payment by credit card. Form PTO-2038 is attached.

FILING FEE AMOUNT ($)

The invention was made by an agency of the United States Government or under a contract with an agency of the United States Government.
☐ No.
☐ Yes, the name of the U.S. Government agency and the Government contract number are: ____

Respectfully submitted,

SIGNATURE _____

TYPED or PRINTED NAME _____

TELEPHONE _____

Date ____

REGISTRATION NO. *(if appropriate)* ____

Docket Number: ____

USE ONLY FOR FILING A PROVISIONAL APPLICATION FOR PATENT

This collection of information is required by 37 CFR 1.51. The information is used by the public to file (and by the PTO to process) a provisional application. Confidentiality is governed by 35 U.S.C. 122 and 37 CFR 1.14. This collection is estimated to take 8 hours to complete, including gathering, preparing, and submitting the complete provisional application to the PTO. Time will vary depending upon the individual case. Any comments on the amount of time you require to complete this form and/or suggestions for reducing this burden, should be sent to the Chief Information Officer, U.S. Patent and Trademark Office, U.S. Department of Commerce, Washington, D.C. 20231. DO NOT SEND FEES OR COMPLETED FORMS TO THIS ADDRESS. SEND TO: Box Provisional Application, Assistant Commissioner for Patents, Washington, D.C. 20231.

PROVISIONAL APPLICATION COVER SHEET
Additional Page

PTO/SB/16 (10-01)
Approved for use through 10/31/2002. OMB 0651-0032
U.S. Patent and Trademark Office; U.S. DEPARTMENT OF COMMERCE
Under the Paperwork Reduction Act of 1995, no persons are required to respond to a collection of information unless it displays a valid OMB control number.

Docket Number	

INVENTOR(S)/APPLICANT(S)		
Given Name (first and middle [if any])	Family or Surname	Residence (City and either State or Foreign Country)

Number _____ of _____

WARNING: Information on this form may become public. Credit card information should not be included on this form. Provide credit card information and authorization on PTO-2038.

LECTURE TWENTY THREE: Risk Management

Remark: This lecture covers an overview of issues that are important in order to minimize risk and litigations in the professional life of an engineer. Perhaps the title of this lecture should read "Risk Management – Professional Liability for Engineers," since we really are just touching the surface as far as the Risk Management concepts in business are concerned. However, that sub-area is under the umbrella of Risk Management, and we would like to maintain the above title for the sake of our overall mosaic of an engineering enterprise that is introduced in the summary of these lecture notes. As a part of this presentation, professional liability insurance is discussed, although the approach of seeking insurance is only a part of the solution for the overall Risk Management strategy.

Special thanks are due the following individuals. Ms. Deanna L. Dietrich, Esq., of the College of Engineering at the University of Wisconsin-Madison for bringing the special resource [1] to the attention of the author. Ms. Monica Adams of *CNA* insurance company, for proposing the above title initially, and putting the author in touch with experts in this area. As well as Mr. Richard B. Garber of Victor O. Schinnerer & Company, Inc. (http://www.schinnerer.com), who is a leading expert as pertains to "Risk Management – Professional Liability for Engineers," and has made a major contribution to the following presentation and has provided us with [2].

Special Resources:

[1] *Lessons in Professional Liability: A notebook for design professionals.* (Formerly titled "Untangling the Web of Professional Liability," by E.B. Howell and R.P. Howell). Monterey, CA: Design Professionals Insurance Company, Security Insurance Company of Hartford, The Connecticut Indemnity Company, 1980.

[2] The Risk Management Committee of the American Institute of Architects (AIA), and Professional Liability Committee of the Professional Engineers in Private Practice Division of the National Society of Professional Engineers (NSPE/PEPP), Richard B. Garber, Ed., *Understanding and Managing Risk: A Guide and Voluntary Education Program for Design Professionals, level 1 and level 2 student booklets*. Chevy Chase, MD: Victor O. Schinnerer & Company, Inc., 1998. [Periodically, this source is revised, but it is used as educational materials continuously. Its reading is a must for anyone interested to pursue this subject from technical point of view.]

Synopsis

Before presenting our discussions based on the literature from insurance and business societies, we give the following explanation of *risk* as control scientists know it. The author is specialized in theory of sensitivity and robustness in dynamic systems, which is a subset of control theory and applied mathematics, and has a lifetime interest in the theoretical foundation of this topic. Setting up the following optimization problem in its many variations and possibilities (deterministically or stochastically) forms the starting point for theoretical study of this subject as is well known to most control and optimization specialists.

- Given a specific dynamic system, find the control strategy, *per se,* that **minimizes** the **maximum error** resulting from a specific effect.

The reason we are presenting this ***"MinMax"*** optimization problem is to reemphasize that just as any such problem which can be set in a number of metrics (mathematical spaces) for measuring various objective functions as well as optimization strategies. To minimize a "Risk," subject to a "maximum error," or a "worse" condition scenario, we must define each of the corresponding variables with its own metric and parameters. Thus, one optimal solution in one setting may not be acceptable in all cases. Now, the structure of the lecture is clear. We start by defining possible risks, then present strategies to reduce or minimize its worse effects – on the financial well being of the interested parties – all from the business point of view.

Introduction

The following presentation covers the basic concepts of risk management, followed by some general remarks on managing the professional liability for engineers. This lecture note centers on presenting guidelines for the avoidance of some possible conflict resulting from not so much as technical aspects of the engineering profession, but rather from non-technical and human aspects of working as an engineer. We hope to elucidate these human aspects of working as an engineer and try to minimize or mange the corresponding and inevitable risks of this profession. All in all, a few good lessons for a productive life.

Concepts on Risk Management

The following materials closely follow the discussion in [2]. All projects are, essentially, investments – resources are allocated in the present in anticipation of favorable outcomes in uncertain futures. What is *risk* and how do we manage it?

The tenth edition of Webster's Collegiate Dictionary defines risk as "possibility of loss or injury." This definition reflects the most common way we view risk's potential materialization in a given situation – as a threat. However, for our purposes, we define risk simply and broadly as the ***probability of an unfavorable outcome.*** Now, what is considered unfavorable is relative to what is expected. The ***distance*** between expectation and outcome is complex and often misunderstood. Here, is an excellent example of how quickly we may get into a quagmire. In the former sentence, we had the word distance, which is what we mean by a metric or a mathematical space. What is the best way to measure this distance? Just using the word complex is not going to answer that, although it flags trouble for us. Also, the term unfavorable is not well defined either. There are infinitely many ways to describe unfavorable, but again we are not going to press those issues herein. All in all, managing risk is a complex issue. It is a MinMax optimization of a dynamic problem, which is often multi-input, multi-output, non-deterministic and nonlinear. Fortunately, the conceptual framework for such a process exits and a number of tools have been developed to support the rational allocation and management of risk in project delivery [2].

Similarly, and before we discuss some technical concepts, Risk Management must be defined. According to [2], this is defined as ***the process of minimizing the probability and severity of an unfavorable outcome at the lowest long-term cost to the organization.*** This process involves three interactive steps: Risk Analysis, Risk Response, and Risk Control.

A. *Risk Analysis* [2]

This topic entails a problem-seeking activity, which involves identifying the sources of risks applicable to the project, assessing their probable impact on the project, and creating a "short list" of the more problematic sources of risk for specific response. In theory, the sources of risk that could potentially impact a project are virtually infinite in number. In practice, the number of statistically significant sources of risk is relatively small. Once the probable sources of risk have been identified, a realistic assessment considers both the probability and potential severity of a given risk event. For example, some problems may occur infrequently, but when they do, the results can be catastrophic. Other problems may occur frequently, but their occurrence is of relatively minor consequence. Intuitively, these situations deserve different responses. Under a given set of circumstances, the *magnitude* of risk is a function of the probability of an unfavorable outcome and the severity of the consequences of that outcome.

B. Risk Response [2]

This step entails the following options: retaining and mitigating the risk; transferring it wholly or partially to another party; or avoiding it completely. It is unrealistic for design professional, or any other project stakeholder, to think that it is possible to avoid all risk. Careful planning to mitigate the risk that has been identified and retained is a good business practice. Currently, there is also much talk about "partnering." Although, that term seems to have multiple meanings, the substance is generally the same. Partnering in its many forms is about communicating and working together to anticipate, identify, and solve problems as quickly as possible. Partnership is an attitude: it requires a commitment on the part of all those involved with the project – owner, design professionals, and contractors – to work toward a successful project and avoid adversarial relationships. A "partnered" project can pay dividends to all participants by reducing the probability of disputes. However, partnering cannot be expected to undo the damage done by an inequitable or inappropriate allocation of risk among project participants.

Another option available for management of risk is to transfer that wholly or partially to another party – say, through insurance or contract. Insurance should not be perceived as a "silver bullet" solution for the management of all risks. Contracts are another vehicle for transfer or allocation of risk. Unfortunately, contracts are a much-abused vehicle for risk transfer. Some clients of design professionals, and contractors, try to allocate all risk to other parties, which is unrealistic. It seems any given risk should be borne by the party best able to control the circumstances creating the risk and best able to insure against the risk. As a general rule, if a party has the power and authority to carry out its duties, it has the best opportunity to minimize the risk associated with those duties. Conversely, parties should not be allocated responsibilities over which they have no control. Sometimes design professionals should consider asking the client to retain or assume an equitable amount of risk. Many times, the real risk is not that the design professional will ultimately be held liable for problems that were not within his/her reasonable control. Rather, the risk is that he/she will be embroiled in litigation about problems that will prove to have been caused by others, but that will sap the time and money of the design professional. Many if not most professional liability claims are predicated, to some degree, on communication failure between the design professional and the client.

Two common ways of addressing these problems are: (1) indemnification, and (2) limitation of liability. These are two different but related concepts.

Indemnification occurs when one party agrees to pay for liabilities incurred by another party.

It is also possible, either alone or in conjunction with indemnification provisions, to negotiate a **limitation of liability** provision for inclusion in the professional services agreement. Under such a provision, one party agrees that it will not seek more than a limited amount of damages from the other party for certain actions or failures to act regardless of the actual amount of the damages. Such an agreement only binds the two parties who have agreed and does not limit a third party's right or ability to recover damages that is due under the law. However, an indemnification in the design professional's favor is not completely adequate for risk allocation or transfer mechanisms. Thus, avoid the risk as much as possible, even when one needs to pass a project.

C. Risk Control [2]

Risk control says that it is not enough to identify the sources of the risk applicable to the project, assess their probable impact, and develop specific response strategies. Those strategies must be implemented and monitored. Furthermore, during project execution, the design professional must analyze and respond to new sources of risk introduced by changed conditions or changes in the design professional's scope of services. Effective control of risk will only be accomplished through continual risk management planning and replanning. Maintaining open and frequent communications and promptly responding to problems during project execution is also essential to the effective control of risk. Avoiding a client complaint or potential liability problem will not cause the problem to go away and, in most situations, will increase the probability of an unfavorable outcome for the design professional and exacerbate the situation.

Legal Liability of Design Professionals

Clearly, rigorous legal discussion is outside the scope of this lecture note and we strongly advise to consult with an attorney should you need one and have additional questions. Thus, the following remarks should be considered as some sort of general information rather than legal advice.

Professional liability consists of those obligations which are or will be legally enforceable, and that arise out of the performance of, or failure to perform, professional services by the design professional [2].

Experience teaches that legal claims against design professionals are usually based on the law of contracts or the law of torts. Torts are civil wrongs, i.e., violations of the personal, business, or property interests of private citizens.

Detailed written contracts for services are more prevalent between design professionals and their clients than in the case with practically any other profession.

As an engineer, whether you are self-employed or work for an organization, you must reduce your vulnerability to any kind of litigation, because even an unfounded lawsuit can damage your reputation and financial well being. The root cause of most misunderstanding is in miscommunication. One must try to minimize that by avoiding failure in interpersonal relationships. It is always to the best interest of all parties involved to avoid litigation, though is not completely avoidable. The insurance policies issued for various forms of professional liability are the last resorts to minimize loss resulting from professional liability, but certainly not the best recourse.

Although professionalism brings respect to our judgments and decisions, we should protect ourselves against possible allegations through some sort of insurance policies, depending on specifics of our activities. For instance, many of us may benefit from a policy on **errors and omissions.** In general, most clients show some understanding and forgiveness when we are conducting ourselves in professional manner. But they may not forgive us always and we should not put our guard down and stop protecting ourselves. All engineers are highly vulnerable to claims from clients, because human nature is very complex and unpredictable. One must be above reproach in every aspect of his/her dealings with clients. However, do not rely on anyone, things change in a blink of an eye. In the following we present some points of views on the importance and procedures of effective communications, contractual words and phrases and other similar practices in order to minimize disputes.

Effective Communication

It is the nature of engineering profession to work as a team comprising diverse personalities from a host of people from around the world. We need to communicate and interface with these people to resolve inevitable conflicts. There is no global set of acceptable rules to define the protocol of proper communicating or etiquette. People from certain parts of the world are more sensitive to certain gestures or ways of communicating than others. Perhaps the best way to open up a dialog is to build trust and there is one magical word, which will do that miracle quite often. That word is the simple first hello, which helps to open the needed trust.

A large number of lawsuits are brought against engineers not for the quality of their technical work, but rather because of the lack of trust and deterioration of personal relationship between engineers and their clients. Particularly, when the initial disputes are not properly handled. These simple and perhaps forgivable errors are then accumulated to bring the relationships to the meltdown stage. As an engineer we need to anticipate and prepare ourselves to deal with possible misunderstanding with our clients. As stated before – engineering is truly a global profession, and one must realize that logics, habits and traits are not globally uniform. Prejudices exist and a simple misunderstanding may snowball itself into many years of litigation. Thus, do whatever it takes to prepare working with all sorts of people. It is a well-known fact among psychologists that most science-based professionals lack sufficient interest in people-oriented vocations. When they are forced to work together the stress comes into surface. Interpersonal skills are critical in dealing effectively with others and this course, we hope to think, has served this purpose to some extent. Effective technical writing that we have emphasized in this course is helpful, but we still need to be effective in our presentation and verbal communication. Words we use quite innocently to describe a situation may have different effects on our audience, because first of all many words have different and often contradictory meanings and we must be certain to convey the right meaning. Secondly, our presentation and body language may give different meaning to a word from the intended written document. Thus, do not assume that you have reached a common understanding with your clients unless they say so. One will be surprised how often these misunderstandings exist without anyone knowing about them.

Contractual Words and Phrases

In almost every country what finally stands in the court of law is a signed contract and even that the corresponding judge may interpret the clauses in the contract his/her way. In setting up a contract do not use extreme words, or words with multiple meanings, or promising words. These types of words leave room for interpretation and therefore encourage litigations. Avoid using dangerous words as much as possible. For instance, consider the following statement.

> We complete our task with the best possible final outcome that covers all facets of this job such that each and every one of its components is in full compliance with all applicable standards of excellence now and in the foreseeable future.

This blank check kills indiscriminately its author and/or signatory. No matter how we look at this sentence, we have dug ourselves deeper and deeper into a situation that can become a nightmare. Thus, modify those clauses in a contract that may become troublesome in a possible court proceeding. Sometimes we may entangle ourselves in our own optimism and use promising words that later add to our risk. These promising words should be curtailed as much as possible. Be pragmatic and anticipate court hearings – thus choose words in a contract very carefully. Ask for expert help.

Do not leave behind a sloppy paper trail that can haunt you. Read the contract before signing and get all your responsibilities in writing. Ask others (at least two) to read your contract/writing and give you feedback before releasing. Especially ask people with experience to read and evaluate critical words used in a contract. There are at least four advantages to being careful about using proper words.

- When you establish yourself as a careful and meticulous reader (first step before good writing is good reading), the people who work with you pay more attention in their correspondence with you.
- When you say what you mean, you eliminate ambiguities and that puts all people around you on notice. If a word creates ambiguities and problems, avoid using it altogether.
- When you write correctly and pay particular attention to basic grammar and spelling, you eliminate errors and portray yourself consistently as a professional.
- The above conduct reduces the risk of defamation suits involving libel or slander and possible technical ambiguities.

Drawing symbols and legends, which are part of a contract should be prepared within the standard of the discipline or explained as clearly as possible to avoid any future misunderstanding. Keep track of symbols, abbreviations and acronyms – particularly in drawings. This way you make the job of other people easier and that is always welcomed.

One area that often leads to litigation is giving an estimate for a task. The client always assumes that estimate means the maximum cost under the worse scenario. Thus, instead of using the word estimate, say "projected minimum cost, which, in our opinion, represents only that, the minimum cost, and the actual cost may be more than the projected." Ask an expert to draft your contract, it pays in the long run.

Conference Discussions

When meeting your clients, be well prepared and keep a complete log of your conference discussions. Send a confirmation and letter of understanding to the client in order to avoid any possible miscommunication. A progress report after that meeting also helps to keep the good faith between you and your client.

Behavior under Stress

It is a well-known fact that people do not function effectively under stress and that may add to creation of risk and subsequent litigation. Avoid decision-making under stress. Being able to control one's stressful condition avoids further escalating of crises. Here are a few suggestions to calm the situation when under stress and thus minimize risk of future litigation.

(1) Remain calm and collected no matter what – remember things can get even worse than this situation.

(2) Do not blame yourself nor assume any guilt – remember you have done as best as you can.

(3) Do not withdraw – remember to quit communicating will not help, always face your adversary.

(4) Ask a confidant to evaluate the situation for you and really sort things out – remember there are experts whose jobs are being intermediaries.

(5) Face your adversary with documents and facts – remember there is always a chance that your adversary also had a bad day and cannot recall all earlier agreements with you.

(6) Remain flexible to alternative solution – remember you may avoid litigation by making some minor changes. "Half a loaf is better than no loaf."

(7) Try to seek mediation by getting help from experts and avoiding expensive litigation through the court systems that are often overworked and biased. There are procedures on seeking help from either *mediation,* which is a sophisticated form of negotiation, made a powerful tool by the participation of a third party called a *mediator,* who facilitates the discussion between you and your adversary. Or *arbitration*, which is less costly, but possibly more efficient thre fair than litigation. The legal structures of these tools are described in law books of different States. Your counsel can describe and advise you should a need arises.

There are many other ways of resolving a conflict, but in the final analysis the best approach is not having one. Conflict resolution, in whatever shape and form, must avoid the ravaging effects of a professional liability lawsuit. Early resolution of any professional conflict is always beneficial. However, when facing the inevitable, make every effort to settle as soon as possible with the help of your professional liability carrier. Judges and other legal professionals may not have taken any engineering course, and certainly they do not think like engineers – remember that! What we have covered thus far are warnings on how to minimize conflict with clients. These are just the tips of the icebergs. There are many more issues to cover that are beyond the scope of this lecture note.

Finally, we particularly warn you to watch out for various contracts that you will be asked to sign when starting a new job. Do not sign any contract – as much as you may love the job – until you have asked a counsel to read it through. Remember, as much as you may want and need the job – so does your employer who needs to fill the vacancy. Take your time and read all papers in your own time and at your own convenience. Big corporations may try to protect their future liabilities by your involvement. So should you – protect yourself by not making a hasty decision under stress.

Recruitment

Although you may not be in a position to hire anyone soon, you need to realize that best organizations treat their new employees very special – not necessarily because they want to – but because in the long run it is to their best interests. The main purpose of this section is to emphasize the other side of the coin and thus teach you a few lessons of life in reverse order!

Hiring staff is a three-step process, whether you do it or it has been done for you. Those are as follows.
(1) Describe the position as accurately as possible.
(2) Find a pool of qualified candidates to meet the job description.
(3) Hire the most qualified person for the job – irrespective of all other possible factors. Remain objective!

Good organizations plan their hiring well in advance, and provide a good orientation seminar for new employees. However, on many occasions these are done at the last minute with little or no preparation to orient the new employees. If that has been your experience in your current job – try to watch for possible risks that are possibly awaiting you, as the result of a sloppy corporate culture. If you are not getting what you were promised – get out as fast as you can. Similarly when you are supervising someone, be aware that if you do not deliver what you have promised you have an unhappy employee who may cause you a professional liability. There is a Persian saying that do not abuse the trust that has been vested on you on the assumption that you are dealing with your friends, because the person that you abused will not remain your friend. Treat people who work with you as nice as you want to be treated yourself – otherwise they will not maintain loyalty towards you.

From time to time, the author sees certain people or organizations boast about their hiring style. In the following we go to the source and express one famous policy to hire staff that is attributed to Dean Everitt's approach (Dean Everitt served University of Illinois at Urbana, College of Engineering). According to Dean Everitt's – the role of the principal person who hires is to locate the most brilliant and creative people that can be found, and provide them with all the necessary resources, then will get out of their way and cheer their successes.

There is also this golden rule of the American business procedure that is practiced by the best in the field continuously. A first-rate person (company) hires a first-rate person, but a second-rate person (company) hires a third- or a fourth-rate person, because a second-rate person does not even want to have someone with capabilities beyond his/hers on the staff.

We hope the above thoughts have given you a good platform to sort the status of your career (when you will start hopefully soon), by reversing the hiring procedures that you went through and realizing where are you standing.

Business Procedures

This is really the heart of your situation, and a major area of discussion. To avoid professional liability, and minimizing your risk, you must conduct your business within the proper standard of State and other local laws and practices. This must be done such that it minimizes the litigation and will provide compliance with the laws. Ask professional accountants and public relation managers to assist you with your business plan. Never ask a relative to do this part for you, especially when you are starting up.

Technical Procedures

Every task comes with a set of specifications, which is the main source of professional liability claims. Try to write your specifications in the contract as clearly as possible – then deliver. Insurance companies issue policies on "Errors and Omissions," but you must be as specific as you possibly can in your contract without leaving out any detail. You should prepare a list of all you need to deliver as a part of your task contract. Recall that the potential for errors and omissions is always present. Therefore to have a good minimum insurance policy gives you a safety net. But the best safety net is your mutual working relationship with your client and your overall attitude of facing a conflict.

Insurance

Professional liability and risk management culminates in an insurance policy that must be set up to meet the major professional liability and possible lawsuits. Many insurance companies issue such policies and often those documents are not easy to understand. The insurance concepts are often complicated. Thus, we must seek help from our broker or the underwriting agent. One needs to know, however, what is the best coverage for the type of the work that one conducts, irrespective of who pays the premium. One must also know how to maintain the premium at a reasonable level, in other words, one will not put the firm under risk of paying more for premiums than it should. In certain instances, one may lose his/her job if the insurance cost exceeds a certain level, even if the employee has not created an actual liability. Thus, it is important to know a few insurance concepts. The following discussion is not complete as far as the insurance industry is concerned, but it is sufficiently generic that it may serve as a good starting point to learn about these topics.

Insurance policies are not easy to comprehend, but every engineer must be cognizant of the contents of his/her professional insurance policy. One needs to know what clauses are in the policy and what they mean. In other words, *one must know which activities are insured and which are not*. Similarly, which activity may increase liabilities and thus add risk to the firm. This information is important also for selecting projects. That means, one will not accept a job if the risk is larger than one can afford. Thus, we must look at the insurance policy as the boundaries of our activities. Also, one must know well that buying an insurance policy is not a science, thus one must shop around for the best coverage at the least price.

There are two categories in most coverage policies: (1) "claims-made;" and (2) "expense within the limit" [1]. In recent policies these names are slightly changed, but the following introductory concepts remain intact. The claims-made policy covers only those claims that are reported to the insurance company during the term of the policy. When the term expires the liabilities expired as well. Policies based on indemnification, however, will pay for tortious actions committed during the policy term even if discovered after the policy expired. The expense within the limit policy is a provision that limits the cost of allocated claims expenses during the term of the policy. Claims-made policies are defined differently by different insurance companies. For instance, "some policies define claims as knowledge of circumstances, which could reasonably be expected to lead to a demand for money or services. Other policies define a claim as a demand for money and services [1]. The second definition means that even though you know someone is likely to ask you for money or services in connection with problems on a project you have worked on, it is not considered a claim until the person actually makes the demand" [1].

"If your policy has the second, and narrower, definition of claims it is important *not* to change underwriters if you know of circumstances that could lead to a claim. If you do, it could lead to loss of your insurance coverage for that incident, because a new insurer always asks about situations for which "you have knowledge" which could lead to a claim. You must wait until either the dispute develops to the point where there is a demand, or you can safely conclude that circumstance will not produce a claim" [1].

"A claims-made policy can, and usually does, apply to claims arising from negligent acts, errors or omissions committed prior to the inception date of the policy. This is called "prior acts coverage." The prior acts coverage usually begin on a specific date, called "retroactive dates," and extends to the inception date of your policy. Claims made for negligent acts errors or omissions committed prior to the "retroactive date" will not be covered. Some claims-made policies provide "discovery periods," which allow the reporting of claims after expiration of the policy if such claims are for negligent acts, errors or omissions committed during or prior to the policy period" [1]. *Each policy is different, it is however important that one buys professional liability insurance continuously to cover both current activity and prior acts completely.*

Expense within the limit policies puts limit on liability that includes most expenses (legal included) and all losses or settlements paid for all claims made under the term of a policy. Thus, if the overall cost exceeds the limit, then that becomes the responsibility of the policyholder. Therefore claims management that avoids unnecessary legal expenses becomes very critical in this approach and should be considered at the outset.

Partial List of Essential Issues Not Included

The lecture note does not address a number of essential risk management processes in project delivery such as pre-project planning, scope management, contract administration, *etc.*. An excellent overview of risk management lecture processes and terms is contained in Chapter 11 of "A Guide to the Project Management Body of Knowledge," which was published by the Project Management Institute, Four Campus Boulevard, Newtown Square, PA 19073-3299, in 1996.

The legal context of professional practice and project delivery is not addressed. A basic understanding of the sources of duty – tort law, statute law, and contract law – is essential to understanding the applicability of insurance in managing risk. Tort negligence is not addressed herein, but is the trigger for professional liability coverage. For additional information refer to Module 1-2 of [2].

The discussion on professional liability insurance is not complete. For more information refer to *CNA*'s Professional Liability and Pollution Incident Liability Insurance Policy, and other pertinent literature, for instance, Module 2-3 of [2].

We close this section with full understanding that what we have covered in this lecture note may serve only to raise the interests of our students in these highly technical matters. There are many other equally important issues in this vast area of risk management, such as bidding, worker compensation insurance, which have not been looked at even briefly.

Final Thought

Insurance policies have two major sets of provisions, which can be categorized into **"insured"** and **"not insured."** The coverage agreement of a typical policy specifies conditions under which the insurance company will indemnify the insured – that is you or your firm. This is due errors, omissions or negligent acts. Thus, the insurance company is at risk on your behalf for damages resulting from your acts. The insurance company may choose to litigate, but often they settle to minimize cost. Professional liability insurers will not insure activity that they cannot measure or quantify. They also shy away from claims that are contrary to the public policy or opinion.

There are also **"exclusions"** sections in a professional liability policy that limits or exclude insurance liability for certain activities. One must negotiate on the scope of exclusion with the underwriters. Insurance companies have their own personnel who evaluate their risk in selling various policies for different professional activities. The key issue is to know what is or is not covered in a given policy.

Make sure that you understand "who is insured," where is insured," and "when is insured?" In other words, is it you, the firm, or any other legal entity that is insured. The place that insurance applies also must be specified. For instance, if someone uses your product in a foreign country, then does insurance cover liability outside of the United States. One must also know that insurance lapses may create problems, so do change of underwriters. Be aware of the time line (before and after) of a policy.

Finally, whatever you wish to do in your life, do it with utmost care and diligence and rely on no one but only on yourself, which is your best protection and the greatest insurance policy. However, you still need a safety net. ***Thus, read and learn your insurance policy thoroughly.***

Closure

The class ends with a great anticipation of the forthcoming design presentation to incorporate all these ideas in that final event and the final course report.

Essential thoughts in this lecture

Issues.	Applicability to your project, if any.
Is your project an asset or a liability?	Analyze as these issues pertain to your case.
Do you want to add anything else?	Please elaborate.

LECTURE TWENTY FOUR: In Preparation for the Final Presentation

Remark: In this lecture we share the rules that we have followed for our final presentations.

Introduction

Our procedures to make the final oral presentations are described as follows. It is suggested to have only *seven* slides (transparencies) to be used for a 15-minute oral presentation, followed by a five-minute video of the actual demonstration in the laboratory, and five minutes for questions and answers, a total of 25 minutes per group. Those groups without videos may use 20 minutes for their oral presentation and demonstration. There will be five minutes break between groups, to set up the video, while the last group is cleaning out their belongings.

In your slides, cover issues that we have described in this course repeatedly. But most importantly, try to present your case as clearly as you can, and with minimum distractions. This is a technical course above and beyond anything else, do not lose sight of that. Do not try to reduce its technical contributions and impacts by not stressing its significance, which may result in degradation in your final grades. Your work must be accurate and complete as far as your problem statement and the meeting of your specification requirements are concerned. Do not exaggerate, nor underestimate your achievement, just describe it precisely and concisely.

The order of presentation is exactly the same as our earlier demonstrations, and in fact is exactly in parallel with the table of contents of our final report. Namely, the first slide should cover the names, and all introductory information about your design project. Use one word, say *"Introduction,"* on the upper right corner of this page to indicate where you are in your presentation. *You may use similar flags in all of your slides.* The second slide covers technical issues, which in turn both problem statement and your design are described here. Continue on the next page to cover the same general technical topics – this is really the heart of your work. All your results must be included by page four, since you are half way into your allowance, and by now should have convinced everyone that you have a genuine design project and outstanding results. Page five should have *a* review of administrative structure of your cooperative work. Budget and/or cost analysis and all such related matters should be included here. The next slide (page six) covers your critical evaluations and design comparisons, meeting various expectations, or otherwise, as described in Chapter Six of your final report. Finally, Conclusions, along the same lines as your Chapter Seven must be described on slide seven. *Faculty will be evaluating these presentations using the same set of questions as in Lecture Note Sixteen, or a similar set prepared by your course instructor.*

Each page may have up to *two footnotes,* to be used for references or similar explanations. *Lessons learned* throughout this experience can be included as possible on all pages, but definitely some on pages two to six.

Final Thought

We have repeatedly used the following remarks in this course. Namely, technical design as conducted by engineers is just *one* tile on the vast mosaic of an engineering enterprise. Again, the following table perhaps will convince us that an engineer must learn both social and technical skills in order to interact with a large group of professionals from an ever diverse host of people from all over the world. To interact with these people, and bring together the product to the market place in a manner to make sufficient income to sustain such an operation, is an art that we must learn, while we are still in school. To educate students about these concepts as they will face in their careers, has been our goal, which we hope we have accomplished. Looking at these matters and understanding the thought behind inquisitively is also a process that we hope has been initiated in our students.

Table 1. The tiles on the mosaic of a typical engineering organization.

Human Resources	Technical Documentation	Marketing & Business Planning	Physical Plants / Facility Management	Risk Management
Engineering Research & Development	Intellectual Property, Patent & Legal	Investors And / Or Shareholders	Purchasing & Procurement	
Safety & Reliability,	Environmental Protection Agency / Other	Banking & Investment	Production And Manufacturing	Advertising
Quality Control	Government Liaison Offices		Accounting & Cash Flow Management	Shipping And Distributions
Ethics	Social Impacts	Public & Customer Relation		Sales

Closure

The class ends with the following remarks. All your logbooks, slides (plus one paper copy), videos and four paper copies of final reports are due on specific time and location as announced. We will also appreciate having two floppy disks of the above. Some of these items will be returned to you for your presentations and according to our announced schedule. Please be there on time and *try to attend at least three more presentations* beside your group. Try to rest and relax, but take nothing for granted during your presentation!

Essential thoughts in this lecture

Issues.	Applicability to your project, if any.
Perhaps this is the most appropriate place to declare that "Each and every piece of these lecture notes forms a tile on the final mosaic of this course. Try to assemble those as best as possible such that it portrays your contributions, and your learning of issues discussed in this course."	Obvious! However, you may still claim that, for instance, your project has nothing to do with certain issues discussed in class. Clearly, and if that is the case, then you must justify your claim.
Anything else have you learned in this course?	Think about it! We believe you have!

Good Luck and Best Wishes!

Part Three: Samples of Student's Work

Part Three: Samples of Student's Work – Introduction

Introduction

In this part we share a few of students' projects that we have been intimately involved in; produced at the University of California at Riverside. This selection by no means implies that those which are not included do not have the same merits, but rather space has been the main factor in this process. The categories of these items are as follows.

I. Mid-Course Report: Here, we have chosen the following report.

[1] Naqeeb Ameeri and Ali Seraj, entitled - ***Motion Detection and Object Tracking in Real Time.*** The Technical Faculty Advisor for this project was Professor Bir Bhanu.

II. Final presentations slides: Here, we have selected four sets.

[1] Christopher Bertell, Hiep Nguyen, Kathryn Wassink, entitled – ***Machine Vision Interface with Seiko Robot Arm.*** The Technical Faculty Advisor for this project was Professor Gerardo Beni.

[2] Kamal Roumani and Jeremiah Kent, entitled – ***Robust Control Testbed.*** The Technical Faculty Advisor for this project was Professor Jay Farrell.

[3] Naqeeb Ameeri and Ali Seraj, entitled - ***Motion Detection and Object Tracking in Real Time.*** The Technical Faculty Advisor for this project was Professor Bir Bhanu.

[4] Karl Doering, Nadim Addus, Heath Deuel, Steven Gyford, Oscar Servin, Michael Wilson, entitled - ***Wireless Communication Systems Block Analysis and Design.*** The Technical Faculty Advisor for this project was the author, in collaboration with Mr. Dave Devries, Mr. Bob Kelly, and Mr. Daniel Kong of the Maxim Corporation.

III. A sample page of entries in the journal of Arun Gupta and Phillip Hoang with some minor editing is added here.

IV. Final Report: Here, we have selected the following report.

[1] Karl Doering, Nadim Addus, Heath Deuel, Steven Gyford, Oscar Servin, Michael Wilson, entitled - ***Wireless Communication Systems Block Analysis and Design.*** The Technical Faculty Advisor for this project was the author, in collaboration with Mr. Dave Devries, Mr. Bob Kelly, and Mr. Daniel Kong of the Maxim Corporation.

Before presenting the above four categories, we share the following personal note with our readers.

Who, when, where and how did it start?

From the right, Keith Cummings, the author, Stuart G. Stanley, Thomas Semetsis, Parviz Lotfi, Warren Teubner, Richard Reed (back), Jee Sook Chung (front), Ikkmo Park, and Danny H. Chung (with the baby) – April 1983. Two senior design projects were concerned with design of special robotics manipulators. The third project, not shown, was concerned with construction of a conveyor.

The origin and the standard of excellence in conducting senior design as well as students' research projects began with the author's affiliation at the State University of New York at Stony Brook during 1979 – 1986, which was a very productive period.

During the academic year 1982 – 1983, the above were the first group of students at State University of New York at Stony Brook, Department of Electrical Engineering, whose senior design projects were considered to be of major accomplishment, receiving special recognition from the University President, Professor John Marburger.

Later, these projects were expanded upon by other extremely talented seniors and graduate students at Stony Brook, lending to the establishment of the **Robotics Teaching and Research Laboratory** – one of the premier and the earliest such educational programs in the country.

I. Mid-Course Report

[1] Naqeeb Ameeri and Ali Seraj, entitled - *Motion Detection and Object Tracking in Real Time.* The Technical Faculty Advisor for this project was Professor Bir Bhanu.

This is an excellent, short and informative mid-course report that was submitted at the beginning of the second quarter for the work performed primarily up to the end of the first quarter.

Midcourse Report of Senior Design Project
Winter/Spring 2001
Professor Mansour Eslami

The University of California at Riverside
College of Engineering
Department of Electrical Engineering

Motion Detection and Object Tracking in Real Time

Prepared by: Naqeeb Ameeri & Ali Seraj

Technical Faculty Advisor: Professor Bir Bhanu

This project is supported by the Visualization and Intelligent Systems Laboratory at the
University of California Riverside

1. Introduction

Background Information

Motion detection has been a key issue for researchers in the field of computer vision and image processing for many years [7]. In particular, computer image processing involving large images of XGA (extended Graphics Array, 1024X768) resolutions have not been possible until recently. Images are stored as large matrices in a computer system's memory. These matrices hold the pixel location, brightness and color value at that location. Images containing 256 different color levels at XGA resolution can become very large, anywhere from 100 kilobytes to 6,291 kilobytes (6.2MB) [6]. Current computer technologies, including memory and processor speed, are now powerful enough to handle complex computation involving matrices. Modern computer systems also have the ability to perform image processing in real time. The phase "real time" has several interpretations. For our purposes, this term is used as a performance characteristic describing the speed at which the images are processed. Using very simple motion extraction techniques we are able to determine the presence of motion. This brings us to our next topic: motion tracking.

1.1 Problem Statement

A Historical Perspective

Once motion has been detected, how can we determine its direction and track its movement? There are several methods that can be utilized. Everyone wants to use the simplest approach, which requires the least resources in terms of time and computing power. The approach should not require the construction of complex electrical hardware and should use existing technology. Preferably, this method should use digital image processing as the means of detection, and software as the means of tracking the object. The simplest approach should also utilize a low-level block operation.

Computers vision tasks break down into a hierarchical of low, medium, and high-level block operations. Low-level operations perform tasks such as edge detection. Edge detection is a common approach for detecting meaningful discontinuities in gray level [3]. Thus, an edge corresponds to a sharp change in brightness of an image. This sharp change in brightness usually corresponds to boundaries between objects. Our project relies on continuously detecting the edge of an object in motion. Thus, the algorithms being used are all low-level operations. There are many edge detection algorithms available but the chosen one for this project is thresholding. On top of that, low-level operations include pixel-by-pixel differencing and mask filtering. These methods are typically the fastest since they involve simple arithmetic operations on matrices.

The next operation in the hierarchical machine vision world is medium-level processing. In this level, edges need to be refined into useful subsets. These subsets can later be grouped together for different purposes such as outlining objects. Our project is not dependent upon subsets and thus, does not utilize medium level processing.

The last of the machine level operations is high level processing. High-level processing involves finding regions within boundaries. At this level of processing, some specific knowledge of the application is often needed. Thus, prior information about an image is needed to recognize objects of interest [1]. Once again, this level of operation is well beyond simple motion detecting and thus, is not utilized. Based on these historical considerations and applications, we offer our problem statement as follows.

> Problem: Given the presence of motion of an object, what is the best way to determine its direction of movement and track the object?

1.2 Problem Solutions

Applicable Solutions

1. Sensor based systems using micro-controllers.
    - Applications: One possible solution to the given problem is the use of motion sensors and micro-controllers. A micro-controller is an integrated circuit special purpose chip used to control other devices. It contains an ALU, memory, and I/O circuits. It is a special-purpose device suitable for a variety of applications. Typically, a micro-controller must respond to inputs within a short period of time to produce outputs [4.1]. The inputs correspond to data coming in from the sensors. The outputs correspond to signals directing the motors to move in the appropriate direction. The sensors could be of various types. One type of sensor is a motion sensor. If 5 motion sensors are laid out to cover a range of 180 degrees for example, each sensor corresponds to an area of 30 degrees. The sensors collectively identify the location of the object [5]. Thus, as an object moves, each corresponding sensor is tripped and this data feeds into the micro-controller, which directs the motors to move a camera to the appropriate position. The advantages and drawbacks are as follows:

    - Drawbacks:
        - Limited coverage area of sensors.
        - In-depth hardware and software interaction, which can be difficult to troubleshoot.
        - Cannot perform image analysis.

    - Advantages:
        - Often inexpensive to design and build.
        - Highly portable.

2. Transmitter/Receiver based system.
   - Applications: A second possible solution to the given problem is the use of a transmitter and receiver based system. In this system, the object would carry a transmitter, which would transmit a signal to 3 receivers. Through simple geometry and mathematics, the objects position can be triangulated [3]. The advantages and disadvantages are as follows:

   - Drawbacks:
     - Object that is being tracked must carry a transmitter.
     - Real World objects cannot be tracked using this approach.
     - Cannot perform image analysis.
     - Receiver range can be limited.

   - Advantages:
     - Object can be tracked regardless of surrounding environment.
     - Does not require complex computer systems to implement.

3. Image processor based system.
   - Applications: A third possible solution to the to the given problem is the use of an image processor. An image processor is a special type of computer dedicated for special image processing functions. The purpose behind this special piece of hardware is that computations can be done a lot faster and often-in real time over a general central processing unit. Algorithms are written for the image processor and are inputted via a serial port on the Sun microcomputer. Once the image processor is done computing, the output is feed back into the Sun through the serial port once again. During this time, the Sun is able to perform other processing tasks. Thus, the use of the image processor helps implement many time dependent functions.

   - Drawbacks:
     - Systems can be very expensive depending on application.
     - Requires complex and powerful software and computer systems.
     - Not easily portable.
     - Complex algorithms must be written in order to extract object movement from surroundings for certain applications.
   - Advantages:
     - Can perform image analysis and extract movement based upon models.
     - Adaptable to most real world situations when motion detection and tracking are required.
     - Software programmable image processor makes changing applications relatively simple.

- These systems are often extremely fast allowing real time processing.

2. Selected Approach

Our selected approach utilizes the image processor based system. We decided to use this system due to the advantages listed above. Also, an image processor is readily available in the Visualization and Intelligent Systems Laboratory at UCR.

Our proposed senior design project involves the use of the MVC 150/40 image-processing computer from Imaging Technologies. The MVC image processor can process images in real time, making it a powerful tool for motion detection, object tracking and various other tasks that most personal computers have difficulty doing in real time. Our project makes use of the MVC 150/40 by detecting and tracking a moving object and by the intervention of command signals to the pan-tilt controller. The pan-tilt controller operates two stepper motors, each responsible for moving the camera mount pan and tilt planes. The controller can interface with a computer through a standard RS-232 serial connection. Commands can be issued through the serial interface and the pan-tilt mount responds accordingly. We use this serial port to issue commands based on the processed images and move the mount accordingly. Thus, the basis for this project involves coding the interface to the pan-tilt unit using the ITEX programming library, as described next.

The ITEX programming library was designed especially for the MVC 150/40. It contains all of the routines and functions that are necessary to use the computer and control the various hardware features exclusive to the MVC 150/40. This includes functions for the onboard serial interface as well as all of the image capture and processing functions. Our design will add to the code, which was designed by previous students. The goal of this project is to develop a suitable program and algorithm that can keep an object centered in the camera's image plane at all times. All of the code that has been previously written for the computer involves motion capture and object tracking only. As of yet, no attempts were made to create an interface for the pan-tilt camera mount. This project establishes a foundation for an automated vehicle control system using the UC Rover.

3. Technical

System Block Diagram:

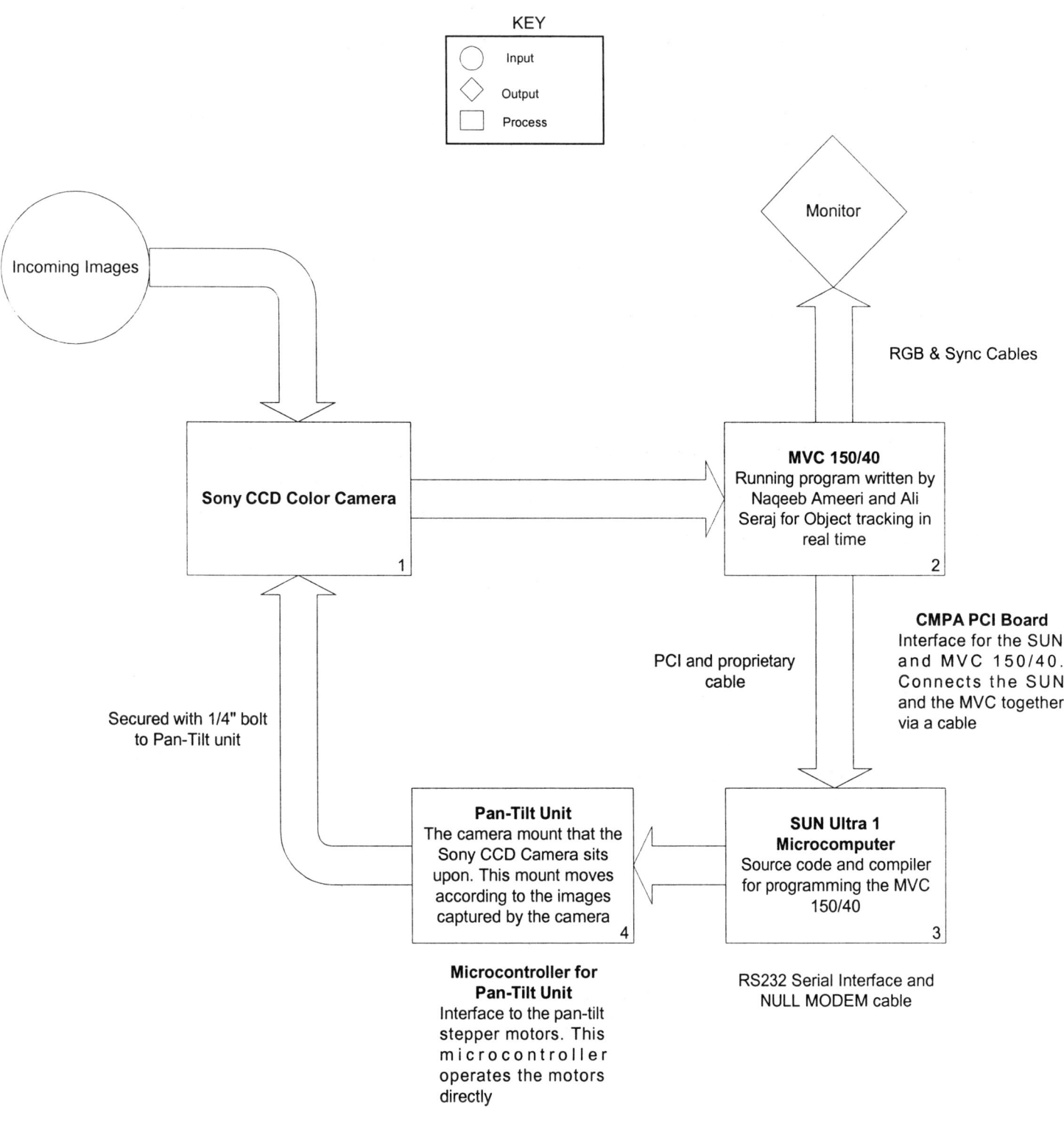

3.1 Analytical Discussions

A. Difference Equation for determining motion:

One of the simplest approaches for detecting changes between two image frames $P_1(i, j)$ and $P_2(i, j)$, is to compare the two images pixel by pixel [4]. If the region of interest allows frames to be captured before and after some time frame, then the pixel-by-pixel subtraction of the two images will yield a picture of what is changed [1]. The subtraction of the two images is performed straightforwardly in a single pass. The output pixels are given by the Difference Equation:

$$Q(i,j) = P_1(i,j) - P_2(i,j)$$

Thus for motion detection, stationary objects cancel each other out while moving objects are highlighted when two images of the same dynamic scene, which have been taken at slightly different times P1 and P2, are subtracted.

B. Lowpass Filtering for noise reduction:

Lowpass filters are used for noise reduction. This is primarily done through blurring. Blurring removes small details from an image such as noise. Pixels that are out of place in a connected cluster are due to noise and must be filtered out [3]. The pixels that belong to an 8-connected component are retained for further analysis. The following 3x3 kernel is used for the non-linear Lowpass filter [4].

$$LPF = 1/16 * \begin{array}{|c|c|c|} \hline 1 & 2 & 1 \\ \hline 2 & 4 & 2 \\ \hline 1 & 2 & 1 \\ \hline \end{array}$$

C. Centroid Detection for a two dimensional image:

After the Difference Equation and Lowpass filter have been applied, the center of the image must be determined in order for a coordinate system representation to be established. The Centroid Detection Algorithm from J. Nguyen and C. Home [4] is used to find the center of an object and is shown here due to its importance.

Centroid Detection Algorithm

1. Initialize variables X, Y, and N to zero
2. Scan the image from left to right and from top to bottom.
3. If the pixel is 1 (black) then,
    3.1 Increment N by 1.

        3.2 Increment X by x coordinate value + 1.
        3.3 Increment Y by y coordinate value + 1.
4. If there are more pixels to consider, then go to step 3.
5. Calculate Centroid: {Xcent,Ycent) = (X/N-1, Y/n-1)

Equipment List:

- **Sun Ultra Sparc 1 Microcomputer**: The Sun Ultra 1 is where the program source code is written and compiled. The Sun Ultra 1 microcomputer is the interface between the programmer and the Modular Vision Computer.
- **Modular Vision Computer** (MVC 150/40): The MVC 150/40 is the bridge between the camera, pan-tilt unit, and the Sun computer. The MVC 150/40 performs all of the real time object tracking and image processing and sends control commands to the pan-tilt unit.
- **Pan-Tilt Unit**: The pan-tilt unit is a 2 axis, 2 degrees of freedom camera mount. The mount is stationary and can pan around 1 axis, and tilt along a second axis. The unit contains two stepper motors each responsible for the pan angle and the tilt angle respectively. Both motors are controlled by an outboard microcontroller.
- **Sony CCD Color Camera**: The CCD camera is the source of images that the MVC 150/40 performs calculations upon.

Software:

    ITEX programming library, GNU gcc ANSI C compiler, GNU Emacs Text editor

Special Resources:

    Visualization and Intelligence Systems Laboratory
       (Marlan and Rosemary Bourns College of Engineering – University of California, Riverside)

4.0 References

[1] R. Boyle, R. Thomas, *Computer Vision: A First Course*. Great Britain: Blackwell Scientific Publications, 1988.

[2] W. Burger, B. Bhanu, *Qualitative Motion Understanding*. Norwell, Massachusetts: Kluwer Academic Publishers, 1992.

[3] R. Gonzalez, R. Woods, *Digital Image Processing*. Reading, Massachusetts: Addison-Wesley Publishing Company, 1993.

[4] J. Nguyen and C. Horne, *Tracking moving objects in Real-Time*, Senior Design Report, Electrical Engineering Dept., University of California Riverside, Riverside, CA, 1999

[5] P. Spasov, *Micro-controller Technology*. Upper Saddle River, New Jersey: Prentice-Hall, Inc.
1992

[6] S. Tzafestas, *Microprocessors in Robotic and Manufacturing Systems*. The Netherlands: Kluwer Academic Publishers, 1991.

[6] D. Vernon, *Machine Vision, Automated Visual Inspection and Robot Vision*. London: Prentice Hall, 1991.

[7] M. Vincze, G. Hager, *Robust Vision for Vision-Based Control of Motion*. Bellingham, Washington: SPIE Optical Engineering Press, 1995.

Part Three: Samples of Student's Work – Final Presentation Slides

II. Final Presentations Slides

[1] Christopher Bertell, Hiep Nguyen, Kathryn Wassink, entitled – ***Machine Vision Interface with Seiko Robot Arm.*** The Technical Faculty Advisor for this project was Professor Gerardo Beni.

[2] Kamal Roumani and Jeremiah Kent, entitled – ***Robust Control Testbed.*** The Technical Faculty Advisor for this project was Professor Jay Farrell.

[3] Naqeeb Ameeri and Ali Seraj, entitled - ***Motion Detection and Object Tracking in Real Time.*** The Technical Faculty Advisor for this project was Professor Bir Bhanu.

[4] Karl Doering, Nadim Addus, Heath Deuel, Steven Gyford, Oscar Servin, Michael Wilson, entitled - ***Wireless Communication Systems Block Analysis and Design.*** The Technical Faculty Advisor for this project was the author, in collaboration with Mr. Dave Devries, Mr. Bob Kelly, and Mr. Daniel Kong of the Maxim Corporation.

The following list shows some of our experience that we would like to share with other course instructors and senior students.

(1) Please pay a particular attention to flags used in these slides, which serve to guide students in the timing and topics of their presentations.
(2) After the experience with using laptop computers and power points earlier in the mid-course presentations, we did not allow that anymore, because students tend to overburden themselves with many graphics and music added to their presentations – their presentations looked more like a made for TV production than actual technical presentations. Occasionally, students changed such presentations without changing their final report, naturally that practice was not allowed in our course. Also, we noted that when one group presents a made for TV presentation, students' concentrations shift completely to outside the classroom. The other groups worried that maybe their technical presentations were not good enough and they lost confidence.
(3) Sometime students using power point tended to show more slides than that they were allowed, a practice not appreciated by the rest of class. By requiring everyone to use the same number of actual "physical" slides (transparencies). Just as any technical presentation requires certain number of pages in a given presentation, *per se*. The entire focus becomes on the substance and actual technical merits of the project and not on the cosmetics.
(4) Furthermore, by asking students to drop their slides in the course instructor's office the night before, they finally go home and come back to present their projects according to the schedule hopefully rested. Otherwise, they want to make changes up to the last minutes before their presentation and that is not a good idea.

We should note that the final presentation is like a thesis defense setting and should be taken very seriously. This is an important overall educational experience for students.

Regarding the fourth set of slides, by Karl Doering, Nadim Addus, Heath Deuel, Steven Gyford, Oscar Servin, Michael Wilson, this is an ensemble of three projects, and as such there is a slight cross presentation among the three groups. This instructor was extremely lucky to have the best senior student of the class 2001 in the electrical engineering program at UCR, Mr. Karl Doering, interested in this topic. He also played the role of technical manager for the group. The business manager of the group was Mr. Michael Wilson who also served as a daily liaison between the author and all team members, though we had regular weekly meetings. This is how senior design projects should be conducted – a marriage between industry and engineering programs.

Introduction

Machine Vision Interface with Seiko Robot Arm

EE 175 Senior Design
Professor Mansour Eslami

Christopher Bertell
Hiep Nguyen
Kathryn Wassink

June 8, 2001

Faculty Advisor: Dr. Gerardo Beni
Technical Advisor: Dan Giles

Background

Problem Statement

To integrate a vision system with the Seiko RT3200 Robot Arm, including electronics interfacing between a PC and the robot arm as well as developing vision demos using C++ or Visual Basic with the Vision System.

Design Specification

To create a system that will use the camera to find and recognize an object, then send commands to the Robot Arm based on the data.

Preparation

Possible Designs

- Hardware
  - Serial *vs.* Parallel Cable Connections
  - Camera Mounting Positions
- Software
  - Microsoft Visual *vs.* Non-Visual Language [VB-RM[†]]
  - C++ *vs.* Basic [MSDN[‡]]

Approach

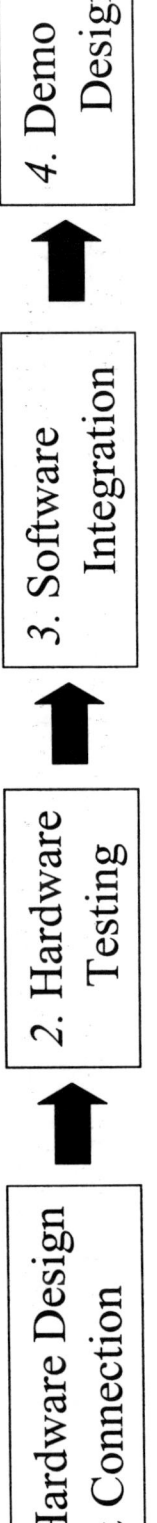

1. Hardware Design & Connection → 2. Hardware Testing → 3. Software Integration → 4. Demo Design

[†] VB-RM: Vision Blox Reference Manual
[‡] MSDN: Microsoft Developer's Network Ocotober '99

The System

Hardware

Camera — Computer — Robot Arm

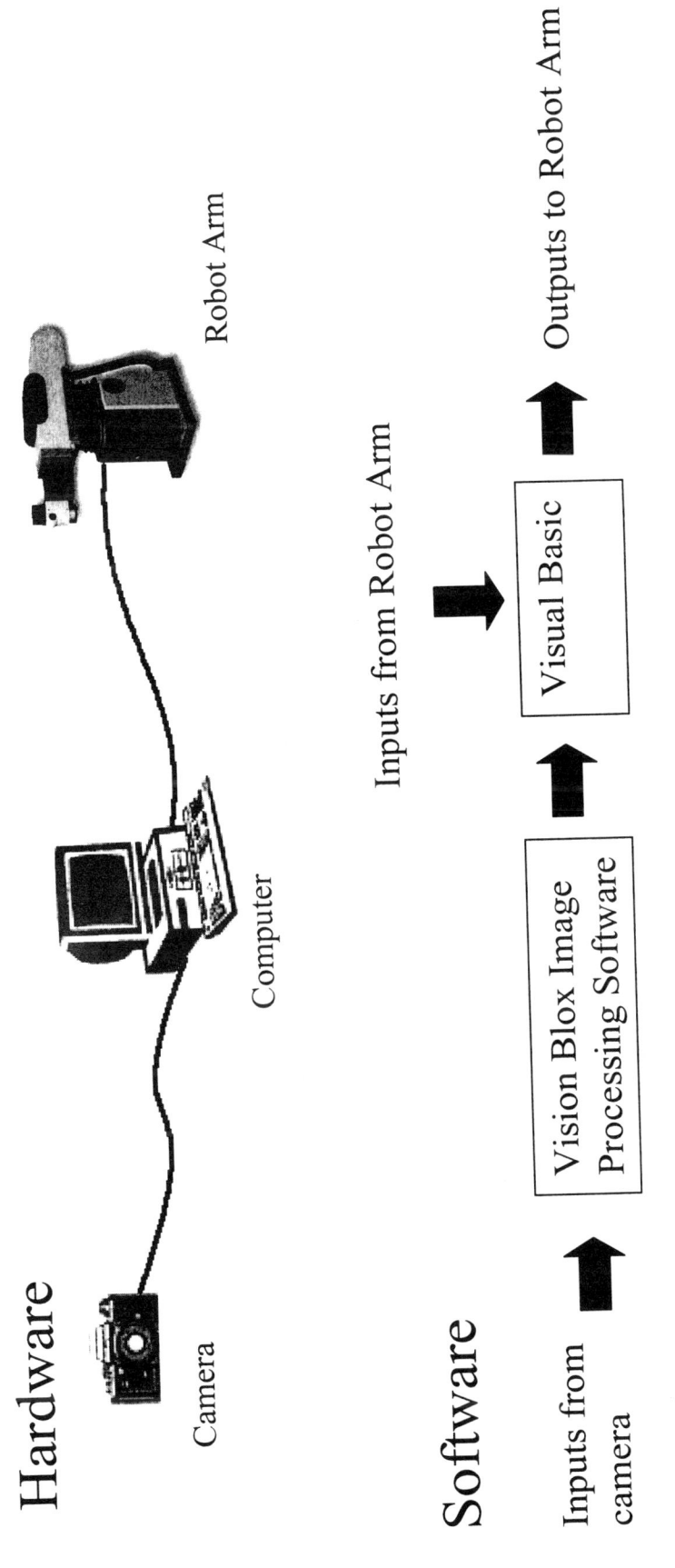

Software

Inputs from camera → Vision Blox Image Processing Software → Visual Basic ← Inputs from Robot Arm → Outputs to Robot Arm

Test Plan

- Use DARL to confirm output to Seiko arm.
- Use PX5 software to confirm camera operation.
- Use demos to confirm software integration.

Critical Evaluations

Meeting Expectations

- Able to create a machine vision interface
- Able to get a camera input and use it to get a robot output

Design Comparisons

- Simpler than original plan
- Cleaner user interface when final product is done.
- Easier to program for serial communications

Elements of Design

- Increases safety by removing workers from potentially harmful situations.
- Probable shifting of workers from assembly-line positions.
- Possible conflicts between workers and companies.
- Improves our understanding of robotics and image processing.

Cost Analysis

- **Marketability**

  It would be an upgrade, primarily for companies with older-model robot arms.

- **Budget**

  Prototype Costs:
  Parts $ 3,025
  Software   220
  Direct Labor 35 hrs @ $25/hr   875
  Capital Equipment   4,000

  Marketing Analysis $ 12,000

  Estimated Costs to Produce 1,000 units:
  Parts (at bulk rates) $ 115,000
  Direct Labor (1.5 hrs/unit @ $25/hr)   37,500
  Storage (1/sq.ft/month, 100 for 6 months)   600

  Advertising Costs: $ 10,000

  Total $ 153,100

  Overhead Cost (85% of above): $ 130,135
  Expected Total Profit $ 90,000
  Grand Total $ 343,235

  Expected Sales   750 units

  Price Per Unit $ 499.99

Conclusions

Administration

- Christopher: Project Leader, Hardware, Control Programming
- Kathryn: Vision Programming, DARL testing
- Hiep: Control Programming, Vision Programming, Hardware

Conclusions

Expansion and Improvements

- New end-effector
- Translation function
- New vision functions: edge detection, rotation, color meter
- New Seiko commands: display, jog, changing speed
- Remount the camera

Lessons Learned

- Don't take anything for granted
- It's all right to ask for help
- Sometimes trial and error is the only way to go

†DARL: **D-TRAN** Assembly Robot Language

Robust Control Testbed

EE 175 Senior Design
University of California, Riverside
Electrical Engineering Department

Kamal Roumani / Jeremiah Kent

June 8, 2001

Professor Mansour Eslami
Project Advisor: Dr. Jay Farrell
Technical Advisor: Dan Giles

Introduction

Technical Issues

■ Current System Problems:
- Not Efficiently Used
- Too Complicated
- Limited
- Too Expensive

■ Design Requirements:
- Experimentally Flexible & Expandable
- Simple Basic Form
- Hands-on Control
- Low cost

Design Specifications

Development

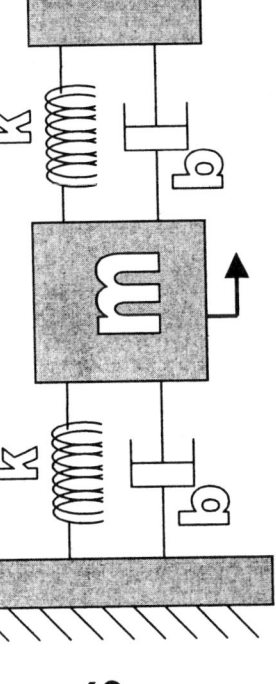

- **Possible Approaches**
  - Tortional, Gantry Crane,
  - Linear, rotational.

- **Our Design**
  - Linear, 1 to 3 DOFs, PWM, DC Motor, Potentiometer Feedback

- **Performance Expectations**
  - Transit 1m track in ~1 second. Low power (3 amp or less).

- **Component Specifications**
  - Motor – torque > 0.325 N-m ; > 550 rpm
  - Power Supply – ±12VDC ; > 1.33 amp

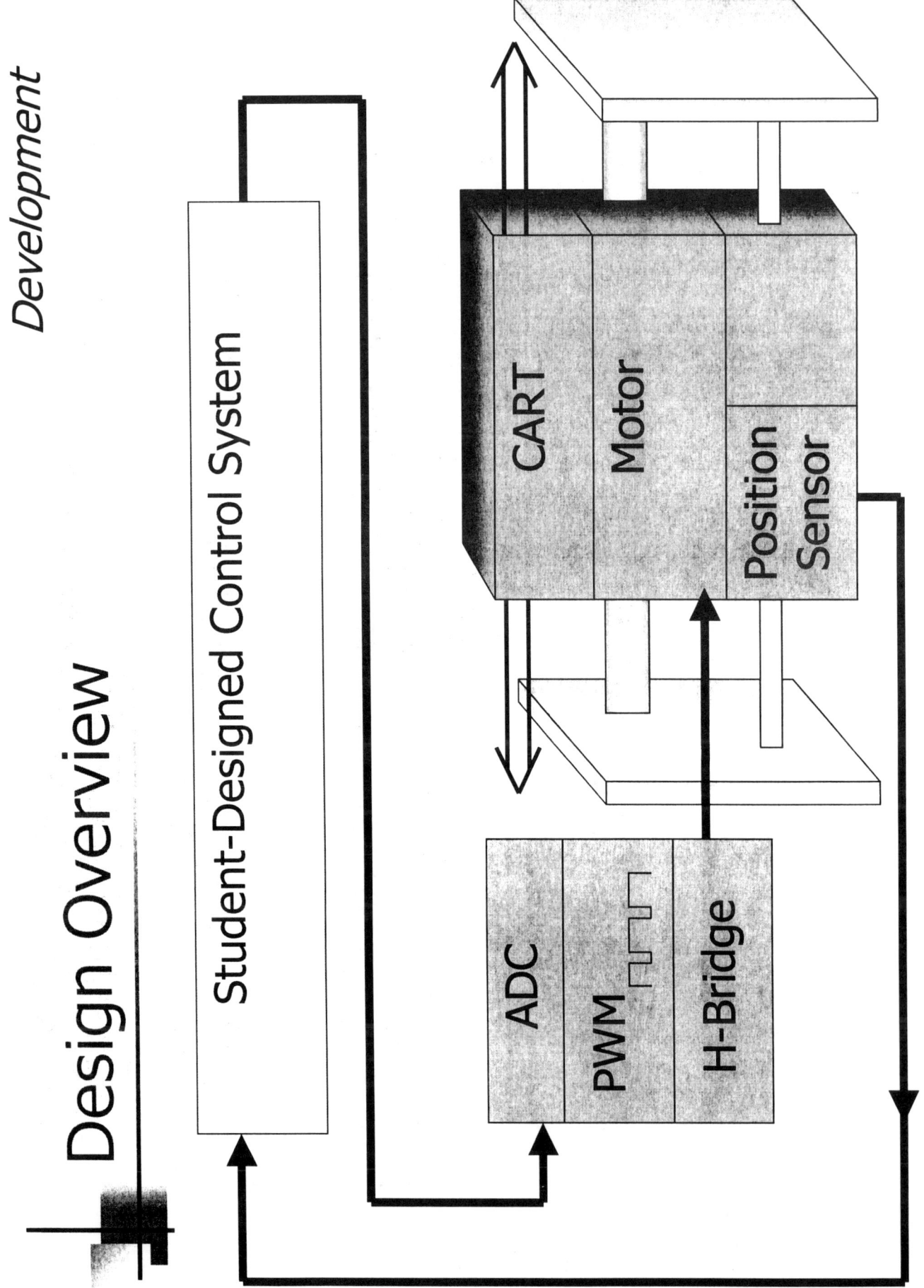

Cost Analysis and Marketability

Administrative

Budget Analysis

- Prototype Cost — $ 10,800.00
- Marketing and Advertising — $ 11,000.00
- Production Costs — $300,000.00
- Overhead Costs — $270,000.00
- Expected Profit — $100,000.00
- Grand Total — $690,000.00
- Expected Sales — 400 units
- Price / Unit — $ 1,749.00

Cost / Unit

Item		
DC motor	$17.50	$15.00
POT	$16.00	$15.00
Bearings	$115.00	$115.00
Gears	$43.00	$40.00
Track	$75.00	$64.00
H-bridge	$5.00	$5.00
Total:	$271.50	$254.00

Results

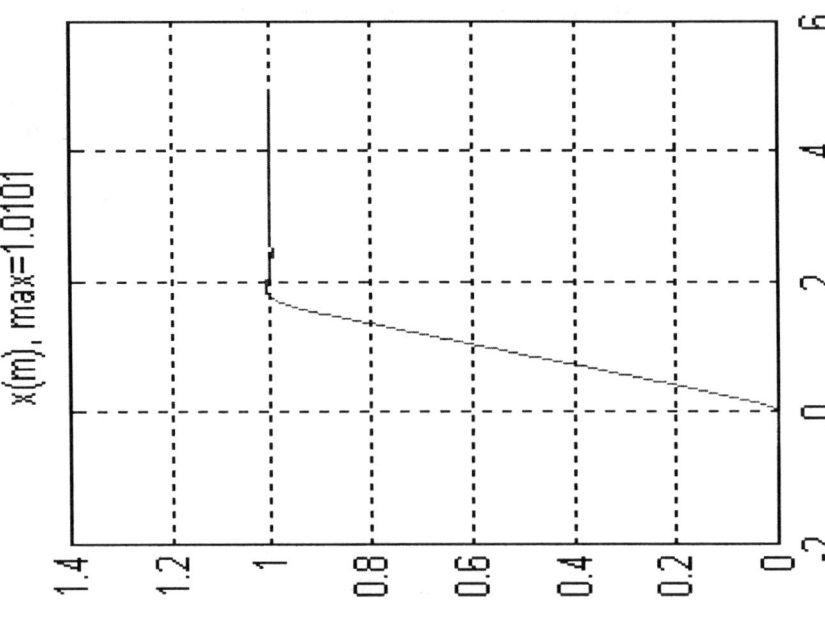

Critical Evaluation

- Due to the motor's internal gearing the damping of the system was significantly higher than originally expected, giving transit times of about 1.5 seconds.

- Ideally a motor with less internal friction would produce a more interesting system to control.

- Our prototype has only one degree of freedom. A production model would have up to 3 degrees of freedom and capabilities for digital control and feedback.

Conclusion

Lessons Learned

- Modeling a system before implementing it can be a Good Thing
- The entire system should work before the first part is purchased.
- No two problems can be solved the same way; be independent thinker and do not use the "rubber stamp" approach (in the spirit of Professor T.J. Higgins)
- **THANKS**
  to Dr. Eslami and Dr. Farrell, as well as to Dan Giles and Mike Fournier

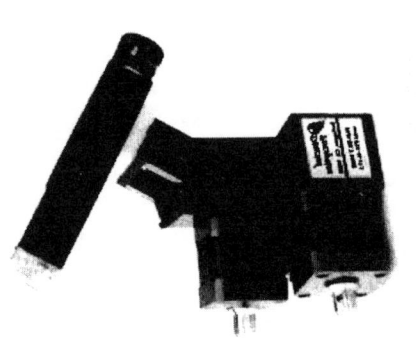

Final Presentation of Senior Design Project for
Winter/Spring 2001
Professor Mansour Eslami

The University of California at Riverside
College of Engineering
Department of Electrical Engineering

Motion Detection and Object Tracking in Real Time

Prepared by: Naqeeb Ameeri & Ali Seraj

Technical Faculty Advisor: Professor Bir Bhanu

This project is supported by the Visualization and Intelligent Systems Laboratory at the University of California Riverside

Introduction

Problem Statement: To implement a motion detection and tracking system in real time.

Applicable Solutions:

- Sensor Based system using infrared motion sensors and a micro-controller
- Transmitter/Receiver based system

Design Solution

Image Processor based system

- **Applications:**
  - Can perform image analysis and extract movement based upon models.
  - Adaptable to most real world situations when motion detection and tracking are required.
  - Software programmable image processor makes changing applications relatively simple.
  - These systems are often extremely fast allowing real time processing.

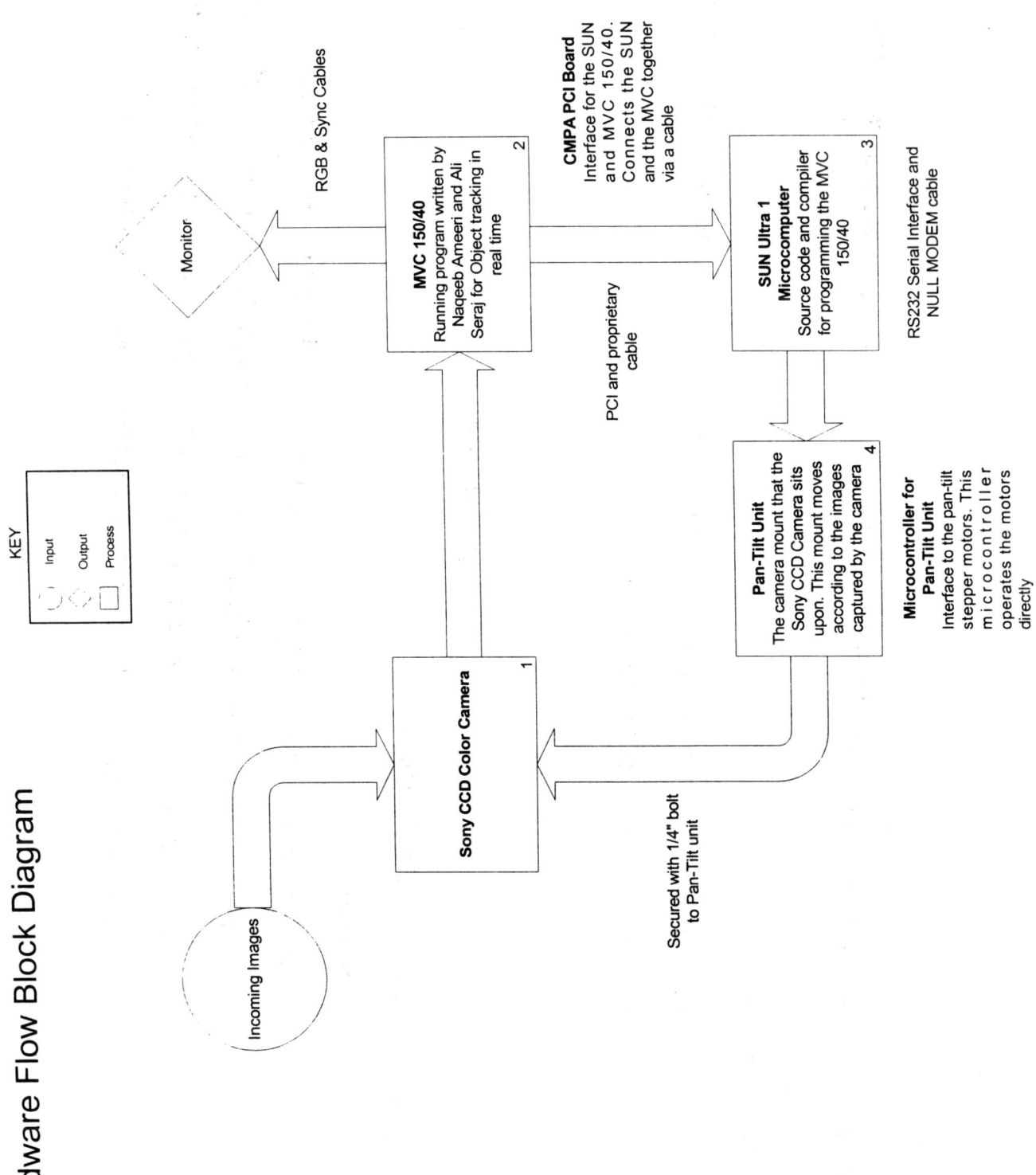

Technical

Design Specifications:

Number of Frames/Buffer/Second   8
Number of Buffers   4
Frame Rate = # of frames * # of buffers   32

Serial Communication:
Serial Interface Type   RS-232 C
Data Transfer Rate   9600 bits/sec
No. of bits   8 data
   1 start
   1 stop
Largest command size (bytes - bits)   64 - 512
Smallest command size (bytes - bits)   32 - 256

Pan and Tilt Unit Specs:
Max Pan Range   280 deg. Left to Right
Max Tilt Range   31 deg. UP
   41 deg. DOWN
Max Pan and Tilt Speed   300 deg./second
Resolution   .05 deg.
Command completion times (average)
   PAN - small (< 30 deg.)   .3 seconds
   PAN - medium (31 to 90 deg.)   .6 seconds
   PAN - large (91 to 140 deg.)   1.2 seconds
   TILT - small (< 16 deg.)   .2 seconds
   TILT - medium (17 to 31 deg.)   .4 seconds

Moving Object
Minimum object dimension (pixels)   35X35
Minimum object speed (change in pixels)   atleast 20

Selected Technical Approach:

Image Processor based system

Input → Difference Equations (f2(x,y,t)-f1(x,y,t)) → Threshold Image Creates a binary (Black White) image → Low Pass Filtering Eliminates Noise → Centroid Detection Determines Objects Coordinates → Output Data To micro-controller

Administrative

Cost Analysis / Business Plan

Prototype Cost:
Parts $6,525.00
Software $0.00
Direct Labor (60 Hours 20/ hour) $1,200.00
Company Facility / Capital Equipment $32,780.00
Depreciation Schedule (40% the first year)

Marketing Analysis:
Expenses for Direct Sales and Advertising Costs $15,000.00

Estimate Costs to Produce 1000 Units: $
Parts (purchased at bulk rate) $6,175,000.00
Labor (3 units/hour) @ 20/hour) $5,000.00
Storage ($1/square feet/month) $2,100.00
Total $6,237,905.00

Overhead Cost (85% of the above) $5,302,319.00

Expected Total Profit $5,620,352.00
Grand Total $17,160,476.00

Expected Sales 750 Units
Price Per Unit $22,880.99

Shipping and Handling $45.00

Schedule

Weeks 20-22: Final Report and Presentation development

Weeks 19: Fine tuning tracking and serial communication

Week 18: Configure software to utilize tilt axis

Weeks 16-17: Configure software to track on pan axis

Weeks 14-15: Enable serial communication between SUN and PTU

Weeks 4-13: Research

Weeks 1-3: Hardware/Software checks

Tasks

What needs to be processed → How is it going to be processed

| Ali | → | Naqeeb |
| Serial Port | | Micro-controller |

Evaluation

Design Constraints

Improving communication between SUN and PTU serial interface

Serial communication speed and command execution time

Computational overhead incurred calculating camera distance from target

Computational overhead incurred calculating background movement and compensation

Elements of Design

Realistic claims based on available data

Motion camera promote public safety

Improves understanding of machine vision and digital image processing

Assists colleagues and co-workers in image processing development

Goals and Design Accomplishments

√ Panning against white background
√ Tilting against white background
√ Tracking an object against an ordinary background

Future Expansion

• The use of a modern image processor and personal computer.
• Background removal.
• Multiple object tracking.

EE 175 Senior Design Project
Professor Mansour Eslami

Maxim W-CDMA Transceiver Design

Senior Design Team:

Nadim Addus
Heath Deuel
Karl Doering
Steven Gyford
Oscar Servin
Michael Wilson

Technical Advisors:

Dave Devries – Bob Kelly – Daniel Kong

Date of Presentation:
June 8, 2001

Project Introduction

- Project Background
  - Welcome professors, fellow classmates, and visitors
  - UCR / Maxim Senior Design Project development
  - Project involves the analysis of and improvement upon a Wideband Code-Division Multiple Access (W-CDMA) transceiver
  - Projects tasks distributed into three major areas of the transceiver
  - Team / Team member introduction

- Communication System Overview
  - Historical perspective
  - Function of a transceiver
  - Various architectures of a transceiver
    - Homodyne
    - Heterodyne
    - Superheterodyne
  - W-CDMA versus other multiple access techniques
  - Critical components / blocks within a W-CDMA transceiver
  - Function of the various components / blocks

Figure 4.2.2.2-1. MAX2360 EV board.

W-CDMA Reference Design

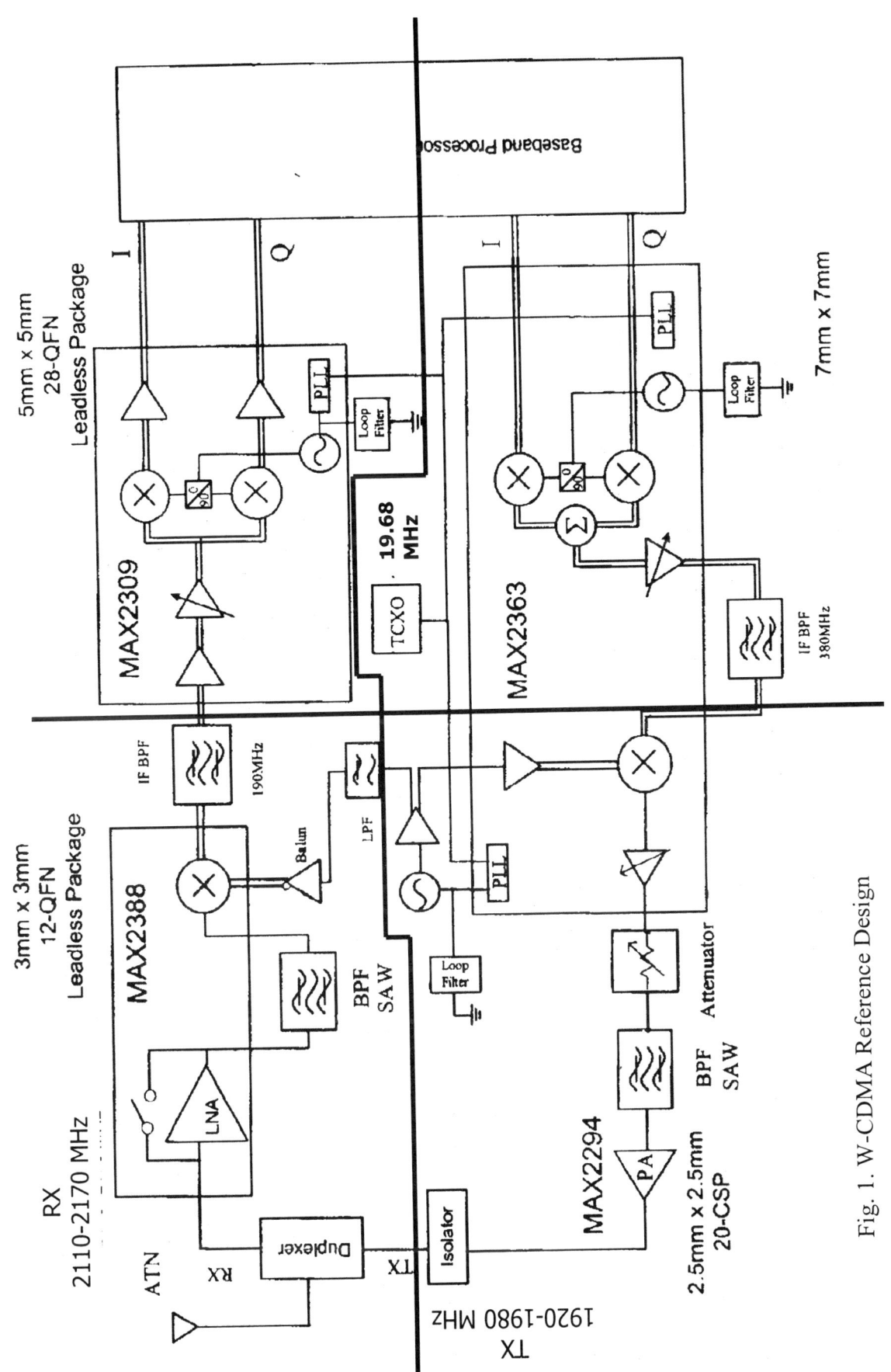

Fig. 1. W-CDMA Reference Design

Introduction

Problem Statement

- Trace signal through various components of the receiver (Rx range: 2110-2170 MHz)
- Calculation and analysis will determine dynamic range, cascade voltage gain, sensitivity, selectivity, cascade noise figure, and cascade third-order intercept (IP_3).
- Verification that reference receiver meets the *Third Generation Partnership Project* (3GPP) technical specifications
- Improvement of adjacent channel interferers rejection of I/Q baseband signals through design of low-pass filter

Design Specifications

Table 2.3.1-1. Required attenuation characteristics at selected frequencies.

Channels	Freq Offset (MHz)	Att. LPF (dBc)
CH-Wanted	0	0.0
CH-Adjacent	± 5	-10.0
CH-In-Band-Blk	± 10	-25.0
CH-In-Band-Blk	± 15	-45.0
CW-Out-Band1-Blk	>± 15	-45.0
CW-Out-Band2-Blk	>± 60	-55.0
CW-Out-Band3-Blk	>± 85	-55.0

Other design parameters:

- Filter output needs to be single-ended
- Order of filter
- Implementation, type, and topology of filter
- Voltage Gain of 6 dB
- Desired signal bandwidth of 1.92 MHz

Method of Solution

Our Design

- Determine order of filter from most stringent constraint; the corner frequency is chosen to be 3.0 MHz
- Filter's resistor and capacitor elements are determined from look-up tables
- Fourth-order Butterworth low-pass filter meets our requirements

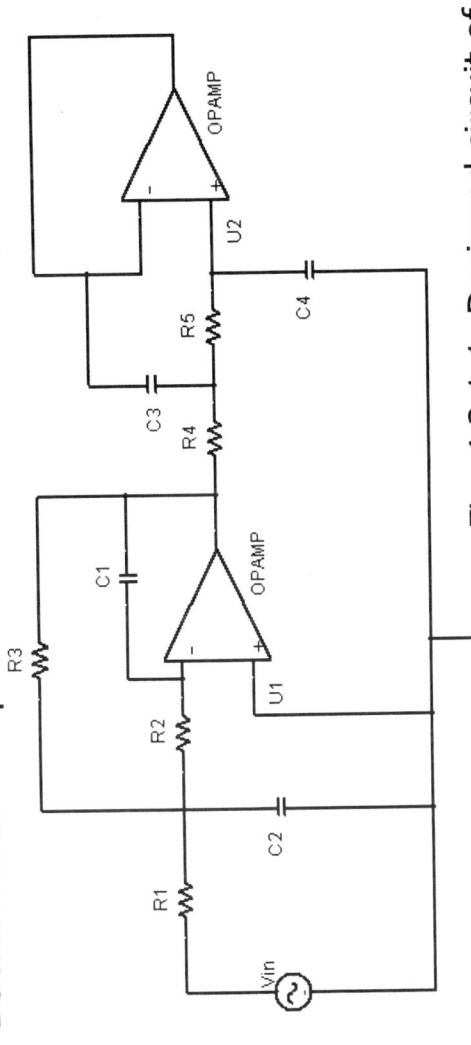

Fig. 4.2.1-1. Designed circuit of fourth-order Butterworth low-pass filter.

Alternative Approach

- **Different filter Implementations:**
    - Chebyshev: Ringing in pass-band causes unwanted noise in desired signal
    - Bessel: Poor attenuation in stop-band requires higher order filter
- **Various filter types:**
    - Passive: Require more space than active filters
    - Digital: Need high processor speed for large bandwidths

Evaluation (1)

Discussion of Results

- **Receiver Performance Summary**
  - Cascade Noise Figure: 2.4 dB which meets 5 dB requirement
  - Cascade Voltage Gain: 12.5 dB
  - Cascade Third-Order Intercept: – 4.12 dBm

- **Filter Performance Summary**
  - –3 dB point at 3 MHz
  - Meets our most stringent requirement of 39 dB of attenuation at 15 MHz
  - Maximally flat response in the pass-band

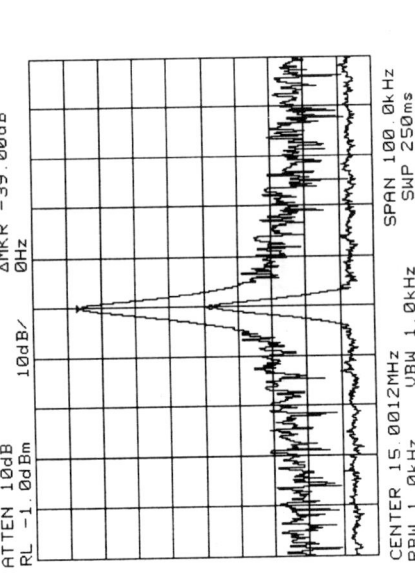

Fig. 4.2.1-4. 15 MHz blocking test of filter.

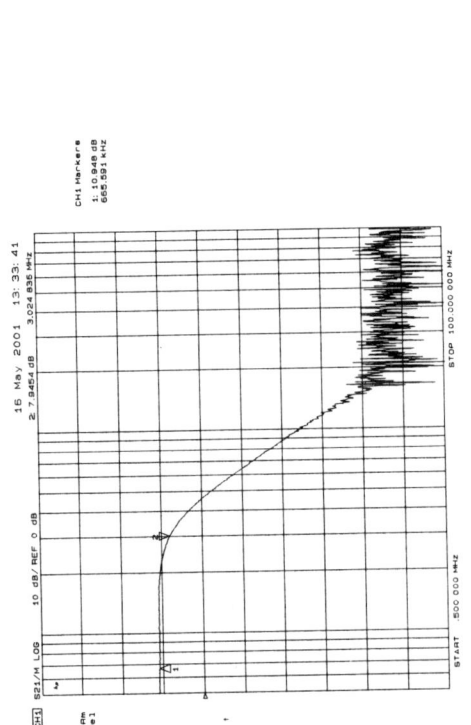

Fig. 4.3.1-1. Magnitude response of low-pass filter.

Evaluation (2)

Design Comparison and Tradeoff

- Simulated frequency response matched actual response
- Group delay (within 100 ns) and phase (as compared to *PSpice* simulation) performed as expected

Category	Simulated	Achieved
Group Delay	49 ns @ 2 MHz	50 ns @ 2 MHz
Phase	90 deg @ 2 MHz	90 deg @ 2 MHz

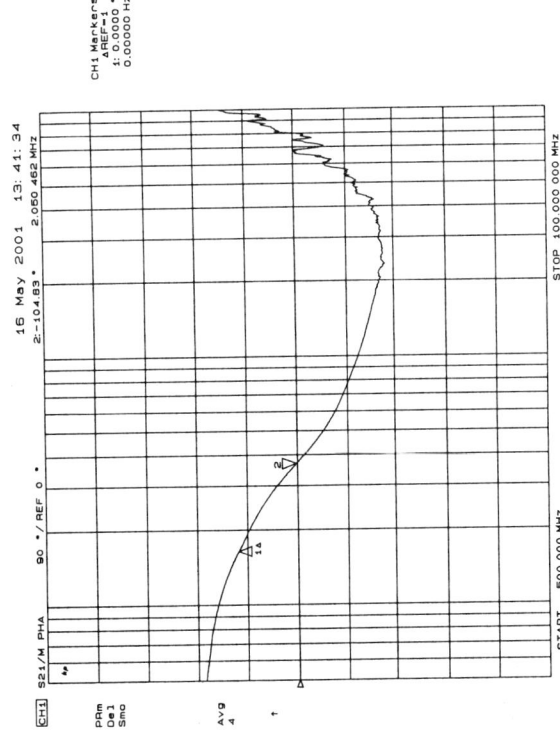

Fig. 4.3.1-6. Phase response of low-pass filter.

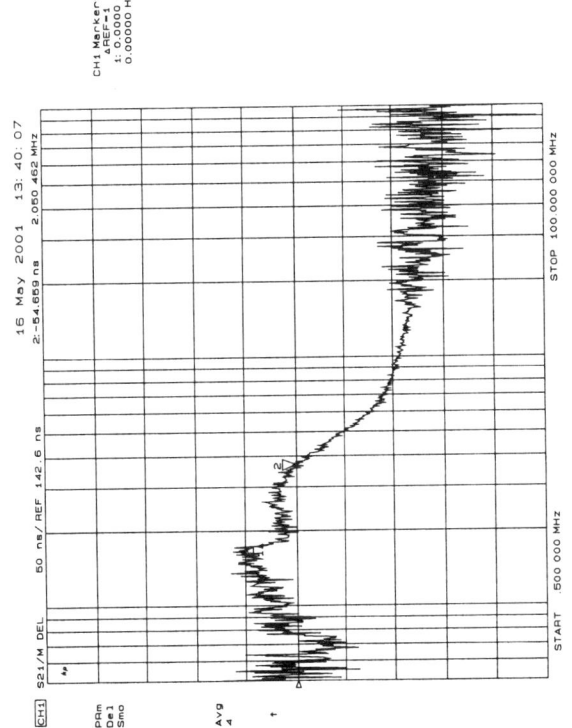

Fig. 4.3.1-5. Group delay of low-pass filter.

Meeting Expectations / Conclusions

Design Constraints
- Lack of component availability of exact designed value
- Pre-existing etched surface mount board
- Op-amp differential output not utilized

Lessons Learned
- Solution broken into more manageable pieces
- Final designed circuit needs additional "tweaking"

Expansion and Improvement
- Temperature considerations
- Design of application specific surface mount board for cleaner circuit

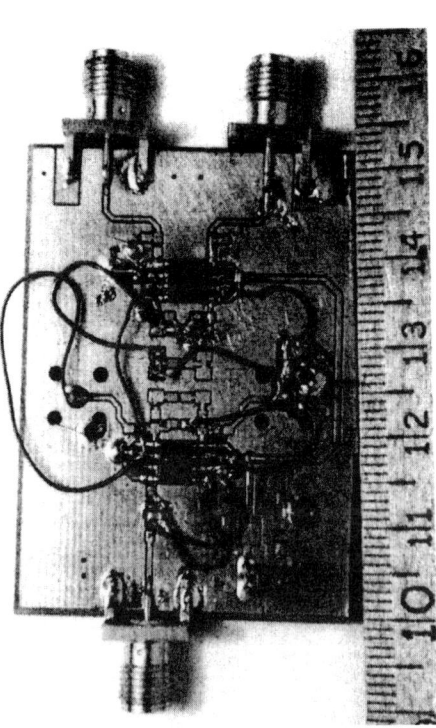

Fig. 4.2.1-2. Picture of actual fourth-order Butterworth low-pass filter.

Introduction

Problem Statement

- The transmitter project is divided into two stages
  - Analysis of the transmitter portion of a W-CDMA transceiver
    - Become familiar with concepts involved in radio frequency design
    - Identify spurious frequency component
  - Design of an intermediate frequency band-pass filter
    - Meet specifications determined by the signal path analysis
    - Implement at lower cost than the current IF / SAW / BPF

Design Specifications

Table 2.3.2-1.1. Transmitter signal path specifications.

Signal Path Analysis Specifications	
Operating Frequency Range	1920 MHz - 1980 MHz
Output Level	23 dBm
Minimum Output Level Range	75 dB

Table 2.3.2-1.2. Transmitter filter specifications.

Filter Design Specifications	
Center Frequency	380 MHz
3 dB Bandwidth (minimum)	6.8 MHz
Attenuation at 190 MHz	37 dB
Attenuation at 760 MHz	40 dB
Differential Input Impedance	400 Ohms
Differential Output Impedance	600 Ohms

Method of Solution

Our Design

- Examine Q_{bp} of filter to determine if it is realizable
  - Determine type and order of filter using steepness factor and tables
  - Initial examinations: 4th order 0.1 dB Chebyshev; 3rd order 6 dB Gaussian; 2nd order Butterworth
  - Selection basis: low-cost, phase response, high attenuation far from f_0, steepness factor
  - Select, determine component values; simulate & adjust for finite Q; transform to differential form;
  - simulate again with available component values

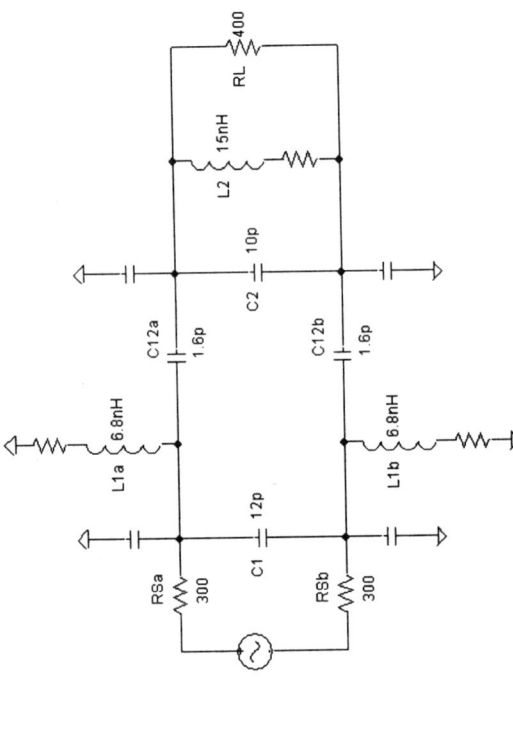

Fig. 3.2.2-2. Simulated response, 190 MHz – 1 GHz.

Fig. 3.2.2-3. Final circuit, after transformation to differential form.

Alternate Approach

- Alternatives: Active vs. passive; wideband vs. narrowband; response characteristics
- Configurations & techniques: LP to BP transform; parallel, series, & synchronously tuned
- Choice of narrowband coupled resonator: high Q, simplified tuning, capacitive coupled

Evaluation (1)

Discussion of Results / Test Plan

- Initial plan: construct prototype on circuit board; measure response; examine effects on TX
- Intermediate trials: construct prototype on MAX2360 EV board instead of unetched board
  - Could not lock PLL at 760 MHz (board specifies an IF of up to only 300 MHz); modified tank circuit
  - PLL locks, but range is too narrow to allow testing of full response of filter (190 – 760 MHz)
  - Bypass on-board PLL with use of signal generator; obtained inconsistent data measuring w/ probe
- Final plan: revert to initial plan using prefabricated PCB with suitable pads and traces
  - Constructed circuit converting differential input and output to single-ended input and output
  - Response was measured on an HP 8753E RF network analyzer
  - Component values were tweaked to compensate for finite Q values and stray capacitance

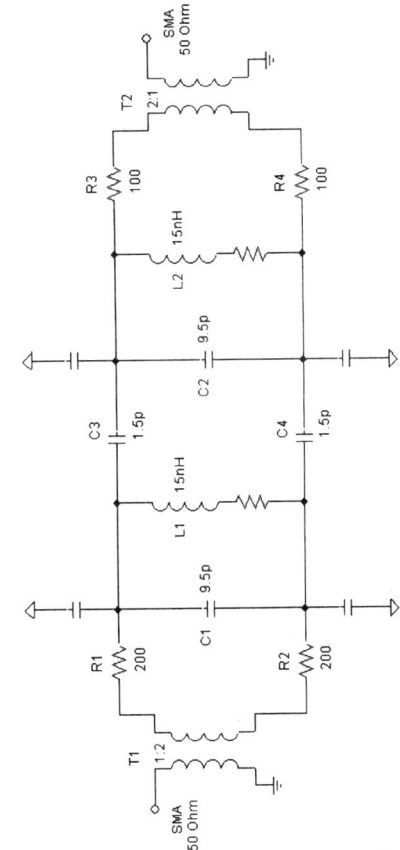

Fig. 4.2.2.3-1. Final prototype circuit.

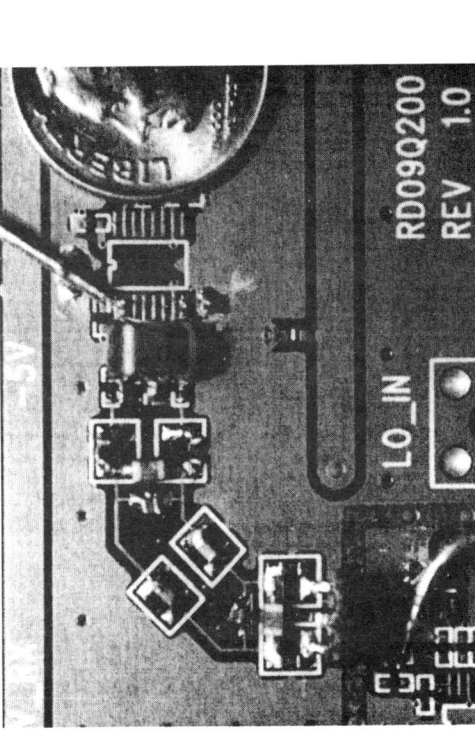

Fig. 4.2.2.3-2. Implementation of prototype (dime shows scale).

Evaluation (2)

Filter Performance: initial response & tuned response

- C_1 and C_2 adjusted several times between 8 and 12 pF to obtain desired response
- Final response of filter: 3 dB BW of 35.4 MHz; insertion loss of approximately 9 dB; attenuation of 39.7 and 44.2 dB at critical frequencies of 190 and 760 MHz, respectively
- Simulated response: 3 dB BW of 41 MHz; insertion loss of 12 dB

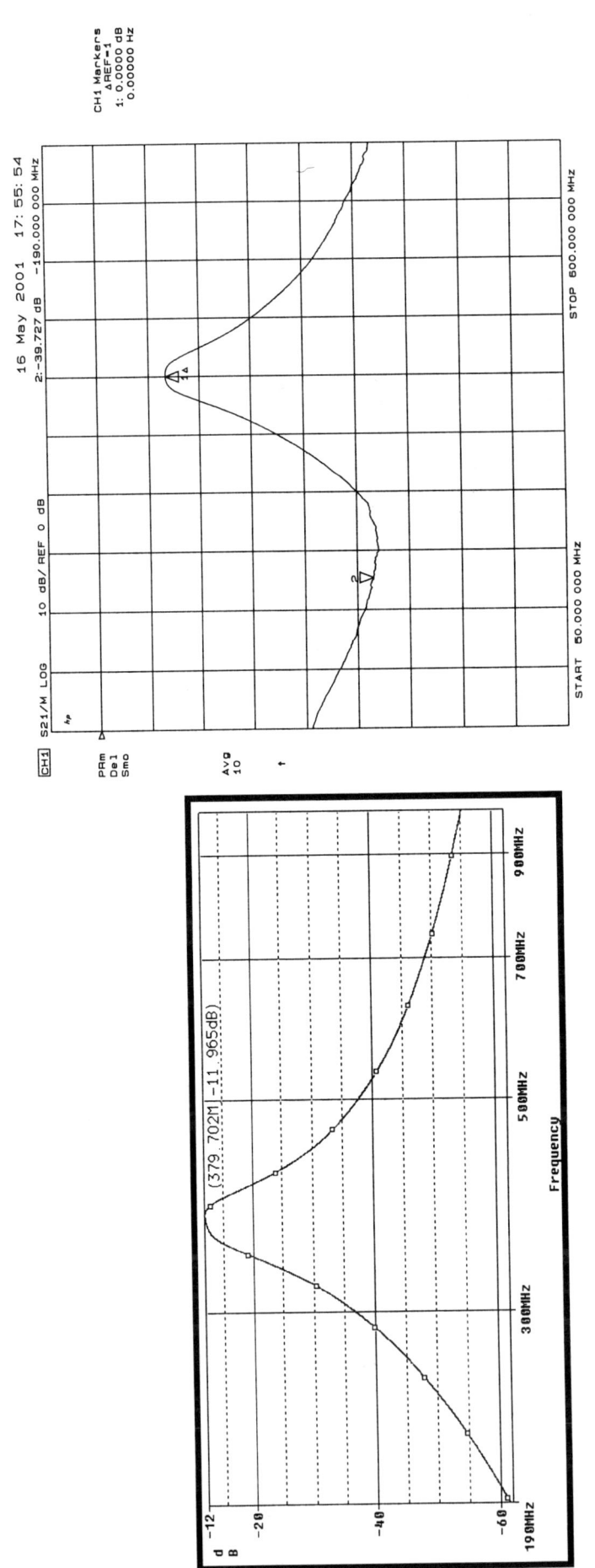

Fig. 4.3.2-3. Measured filter response, 50 – 600 MHz.

Fig. 4.3.2-5. Simulated response, 190 MHz – 1 GHz.

Meeting Expectations / Conclusions

Design Constraints
- Finite inductor Q; stray capacitance; discrete valued components; financial limitations

Lessons Learned
- Design of a low-order BPF at high frequencies is particularly challenging
- Many of our trials contained unrealistic component values, were extremely sensitive to component variation, or did not account properly for finite inductor Q

Expansion and Improvement
- Examine the quadrature modulator, VGA, or power amplifier in the TX path
- Implement the IF BPF on-board a working W-CDMA phone to determine the practical advantages and disadvantages in an uncontrolled environment
- Future senior design teams will benefit from an early start on researching RF basics prior to the beginning of their project

Conclusion
- Filter design specifications have satisfied all technical requirements under the given constraints
- The UCR / Maxim Senior Design Project has been a very valuable learning experience by providing the team "real world" engineering problem solving techniques and has helped establish networking foundations in the corporate environment.

Introduction

Problem Statement
- Four clean Local Oscillators (LO) are needed for down- and up-conversion of our Radio and Intermediate Frequency signals in the ranges from 2360 MHz to 380 MHz.

Solution
- An Integer-N Phase Locked Loop design will be used to tune a Voltage Controlled Oscillator that will supply the transceiver with the proper LO.
- A well designed loop filter is the key to provide a clean signal. Our loop filter is a third-order lead-lag passive low-pass filter within our Phase Locked Loop.

Goals
- Determine the main criteria behind each of the Phase Locked Loops / Local Oscillators.
- Design the control loop components and parameters given some specifications for the radio operation.
- Justify the frequency plan and understand how it impacts the radio.

Method of Solution

Components	Pros	Cons
• Integer-N Phase Locked Loop	• Low Power • Small Circuitry	• High division ratio – significant degradation in phase noise performance
• Fractional-N Phase Locked Loop	• Improves phase noise • Improves loop bandwidth	• More circuitry • Adds spurious noise
• Crystal oscillator	• Clean sinusoidal signal	• Natural resonant frequency way below 100 MHz
• LC Oscillators	• None	• Impossible to design stable and accurate at such high frequency
• Phase Frequency Detector (PFD)	• Guarantees PLL to lock regardless of loop filter	• Adds noise
• EXOR Gates	• Good phase tracking for symmetrical signals (square wave)	• Poor performance when signal is not symmetrical
• JK flip flops (JK/ff)	• Symmetry of signal is irrelevant since JK/ff work on rising edge	• Pull-in range stays severely limited (No integrating term)
• Low-Pass Filter	• Simple, low cost, low phase noise	• Adds noise
• Voltage Control Oscillator	• Tuned to very high frequencies	• Tendency to drift

Design Specifications

INTEGER-N ARCHITECTURE PHASE-LOCKED LOOP

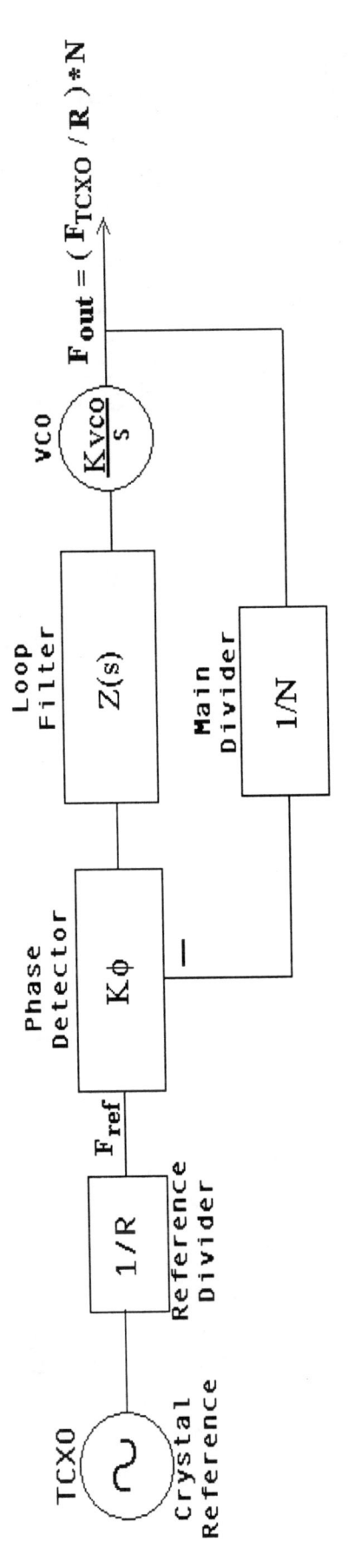

Loop Specifications

TCXO = 19.68MHz   Kvco = 15MHz/V

F_{ref} = 75kHz $K\phi$ = 2.1mA

F_{out} = 760MHz   R = 262.4

F_a = 500Hz   N = 10133.3

T_s = 5ms

Transfer Function

Forward loop Gain = $K\phi Z(s) Kvco/s = G(s)$

Closed Loop Gain = $\dfrac{G(s)}{1+G(s)/N}$

$$Z(s) = \dfrac{s(C2 \cdot R2)+1}{s^2(C1 \cdot C2 \cdot R2)+s(C1+C2)}$$

$F_{out} = (F_{TCXO}/R) \cdot N$

Simulation vs. Experiment

Simulation

K_{vco} = 15 MHz/V
K_ϕ = 2.1 mA
F_{ref} = 75 kHz
Bandwidth = 7.85 kHz
Phase Margin = 70°

Experiment

K_{vco} = 30 MHz/V
K_ϕ = 175 µA
F_{ref} = 75 kHz
Bandwidth = 1.5 kHz
Phase Margin = 78°

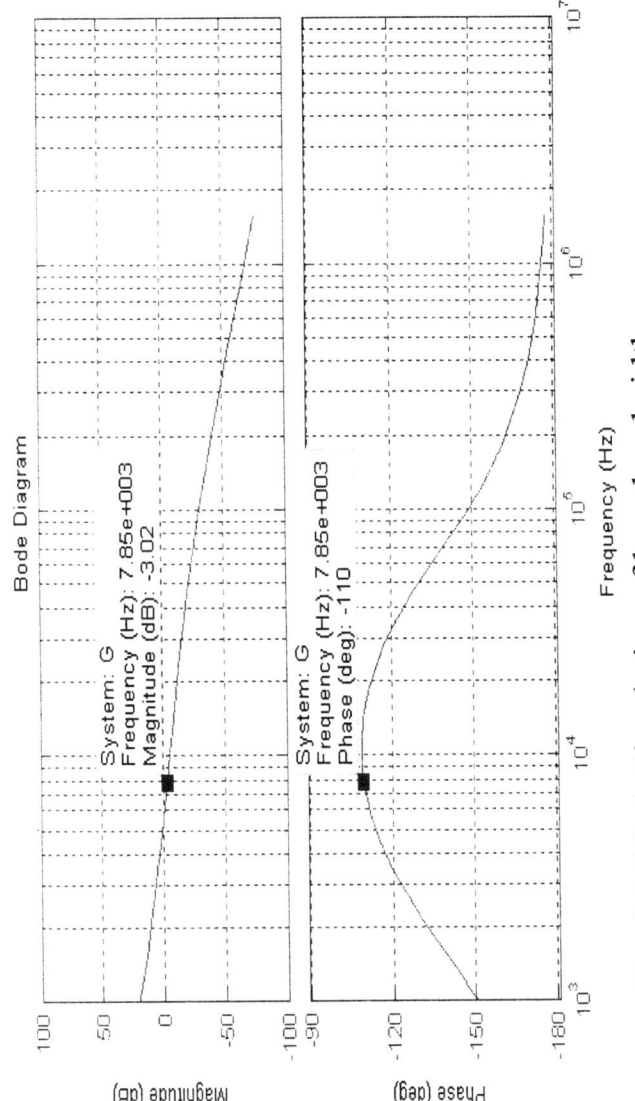

Fig. 4.2.3.2-6. Simulation of loop bandwidth.

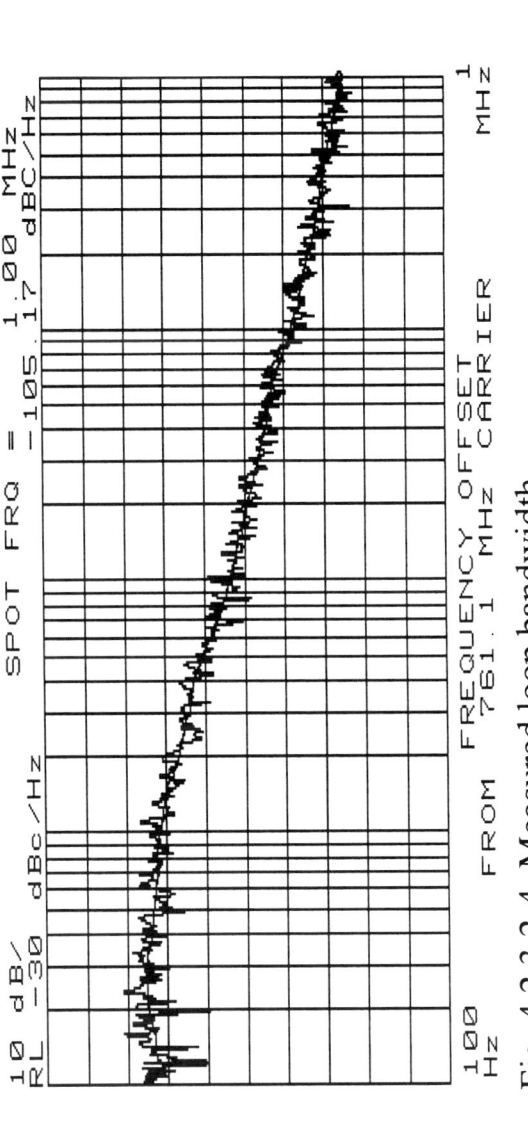

Fig. 4.2.3.2-4. Measured loop bandwidth.

Meeting Expectations

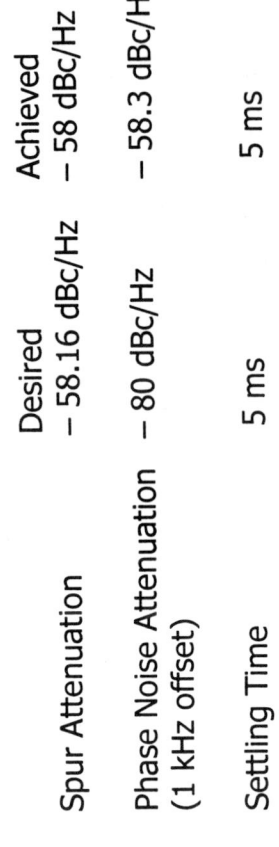

Fig. 4.2.3.2-3. Plot of noise measurements at different offset frequencies.

	Desired	Achieved
Spur Attenuation	−58.16 dBc/Hz	−58 dBc/Hz
Phase Noise Attenuation (1 kHz offset)	−80 dBc/Hz	−58.3 dBc/Hz
Settling Time	5 ms	5 ms

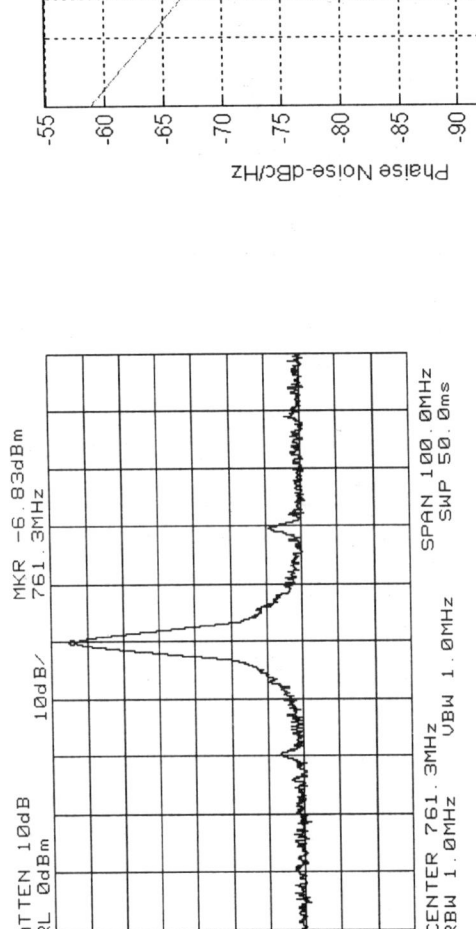

Fig. 4.2.3.2-1. Locked PLL with center at 761.3 MHz and spurs at multiples of 19.68 MHz.

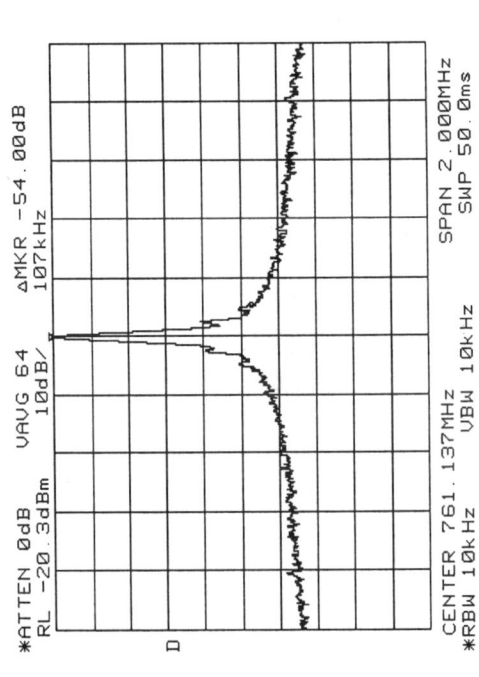

Fig. 4.2.3.2-2. Close-in phase noise.

Conclusion

Mission Accomplished

- After a very exciting 6 months, we have successfully accomplished our mission.

Future Improvements

- Pre-developed time line that gives explicit tasks and dates for which each task should be completed
- Indefinite access to the proper test equipment
- Implementing the third-order loop filter
- Making charge pump current a point of concentration so that power saving techniques can be studied and implemented

Lessons Learned

- Selection of the proper components can have a serious effects on the system
- Part of a design is not only understanding the concepts, but also dealing with non-ideal properties, which adds additional constrains in our design that we didn't necessarily take into account when running the simulation.

Elements of Design

Issue	Effects of our design
Economic Factors	There are many economic factors that influence a design. We have covered a cost analysis in Chapter Five, which gives us a good idea of what it would cost to mass-produce our design. Though this is a good rough estimate, the fact that there are other companies out there wanting to make money from producing a similar product is another factor that must be considered. Through supply and demand, many companies may not be competitive enough to survive and this would cause loss of jobs, however it would also supply the customer with superior products.
Safety	While the actual manufacturing processes involved with mass-producing our design may involve some drawbacks to the environment, all responsibilities fall on the manufacturer to conduct itself accordingly. As for our responsibilities when making a prototype, all known precautions were taken to ensure the safety of all involved.
Reliability	No extensive tests were made to ensure the long-term reliability of our design. However, to the best of our knowledge, we have no reason to believe that our design would not be worthy of long-term use.
Aesthetics	The prototypes that were made during this project were constructed for test purposes only and were not optimized for appearance. They were merely meant to provide us with knowledge and skills related to our particular design. With further design and testing a prototype worthy of being produced, which not only works to designed specifications, but also looks good, could be achieved.
Ethics	In reviewing the IEEE Code of Ethics, we have done nothing but try to better our understanding of this technology as well as help each other along the way, working together as a team. We often criticized and complimented each other during our time working together. Throughout this project, we have conducted ourselves in nothing but a professional manner and have done nothing to harm the reputations of others involved in this project. By reporting all of our work and findings in this report, we have given credit where needed and we have presented our findings with the utmost honesty.
Social Impacts	With many people desiring portable communications devices in today's fast-paced world, our design will play a major role in furthering the wants and needs of these types of people.

Cost Analysis / Marketability

- Prototype Cost:
  - Parts — $530
  - Software — $2,920
  - Direct Labor — $3,600
  - Capital Equipment — $0

- Production Cost of 500,000 Units:
  - Parts — $796,360
- Cost (sub-total) — $803,410
- Overhead Cost — $40,170
- TOTAL COST — $843,580

- Expected Sales — 490,000 units
- Estimated Price/Unit — $1.72
- Current Price Per Unit — $2.00
- Savings Per Unit — $0.28
- TOTAL SAVINGS — $137,200
- Hiring Bonus Cost — $60,000
- **Expected Total Profit** — **$77,200**

Acknowledgements

- We would like to give special thanks to the following extremely helpful people:
  - Dave Devries, Maxim Corporation
  - Warren Hauff, Maxim Corporation
  - Gary Huang, Maxim Corporation
  - Bob Kelly, Maxim Corporation
  - Daniel Kong, Maxim Corporation
  - Kamal Roumani, fellow UC Riverside student who recovered damaged files
  - Leo Valdez, Maxim Corporation
  - Zhongmin Yu, Maxim Corporation

- We greatly appreciate Maxim Corporation for allowing us to use their facilities and resources. We are especially thankful for the help we obtained from each and every one of the engineers, technicians, and support staff who contributed to the success of this project.

- Finally, we thank our faculty advisor Professor Mansour Eslami who made this entire project possible.

Part Three: Samples of Student's Work – A Sample Page of Journal   Page 219

III. A sample page of entries in the journal of Arun Gupta and Phillip Hoang with some minor editing is added here.

Please note that the actual construction of these pages (issues regarding margins, page numbering, style, *etc.*) is described in Lecture Four, which we do not repeat.

Week #19
---

80. (The 80th entry in the Notebook and this was the creation of students) June 1, 2001 (2:00PM to 3:00PM)
We met with Dr. Eslami to discuss issues with the project such as the documentation and presentation. The results are given below:
- The presentation is 7 slides, format is the same as seven chapters of report.
- Add a flag to each page on the upper right of the slide (smaller than large text).
- For administrative, add how we shared the duties, task organization.
- w/video --> 15 minutes, w/o video --> 20 minutes.
- Dr. Eslami also had a chance to look over our documentation and give us pointers on where to improve it.

     Length (hours): 1.0 (Arun)   1.0 (Phillip)

79. June 1, 2001(1:30PM to 2:00PM)
We met with our advisor to discuss all the various parts of the project. Several items were discussed:
- The code provided by us was deemed to be excellent
- We should add low cost to our main problem statement for the CS8900
- Make a point as to the elimination of the PC
- Discuss other solutions, such as direct connection to the PC
- Show how it's too expensive, not standard, too complex
- Add line to the picture for the presentation to show distance
- Show the software stack
- Make the picture crystal clear
- Add the LCD, temperature interface code

Overall, our advisor was pleased with our work.

     Length (hours): 0.5 (Arun)   0.5 (Phillip)

77. May 31, 2001(1:00PM to 5:00PM) *[Note that we skip here and later.]*
We met in the labs to burn the Intel 8051 chip and test it to make sure it works in the lab. We also cleaned up the code for the programs so that we can append it to the report.

     Length (hours): 4.0 (Arun)   4.0 (Phillip)

74. May 29, 2001(2:00PM to 3:00PM)
Arun contacted an advertising agency to determine costs for marketing a product. He consulted jsm+ communications located in Santa Monica California. One of the executives developed the numbers for our business plan as a part of our final report. The details of the conversation are omitted.

     Length (hours): 1.0 (Arun)

  Final Remark: The above trend continued for the rest of the week. The entire history of this project was tabulated in this manner, with complete details on how the thoughts were evolved and how the works were shared during 20 weeks. One of the interesting features of this journal was that they continuously gave the total amounts of time they put into this project, although a further breakdown of that number into different categories of work would have been more instructive. They gave schematics of their designs and their lessons learned, and all the ways that they made corrections and changes week after week. They were candid and open about their experience. They just followed the instruction and it was a joy to work with them.

IV. Final Report

[1] Karl Doering, Nadim Addus, Heath Deuel, Steven Gyford, Oscar Servin, Michael Wilson, entitled - ***Wireless Communication Systems Block Analysis and Design.*** The Technical Faculty Advisor for this project was the author, in collaboration with Mr. Dave Devries, Mr. Bob Kelly, and Mr. Daniel Kong of the Maxim Corporation.

We have selected this item, because of our personal involvement in its creation. We hope that this report would be of both interest and help to our future students.

More importantly, we have included this project to emphasize that a multi-team project must be assembled coherently. As stated in the opening remarks of Section II of this part, senior design projects should be the result of a marriage between engineering programs and industry, a prospect that is fruitful for all parties involved. But that is not an easy task to arrange, since people in industry are under different sets of constraints from their counterparts in academia. In some instances, it may require decades of pursuing and working relationships with industry to build enough trust to let our students in their facility, nevertheless it can be done and should be explored.

Senior design projects may provide golden opportunities for a qualified faculty member to develop and operate a successful research enterprise continually, provided that projects are selected coherently and with an eye to the future. Thus, those who shy away from contributing to this activity are missing a remarkable chance to enhance their research productivity while helping undergraduate students. This instructor has enjoyed the experience of working with many of these young students specially when they come up with their "awesome" results.

Just as any good story, there is a time that one must say farewell and leave the reader. We close by saying that it has been truly an enjoyable year to work on this project. Best wishes to all you young people and so long until we meet again.

Final Report of Senior Design Project
Winter / Spring 2001
Professor Mansour Eslami

The University of California at Riverside
Department of Electrical Engineering

Wireless Communication Systems Block Analysis and Design

Prepared by: Karl Doering,
Nadim Addus, Heath Deuel, Steven Gyford, Oscar Servin, Michael Wilson

Technical Advisors:
Dr. Mansour Eslami, Faculty Advisor
Mr. Dave Devries, Manager, Strategic Applications Engineering at Maxim Corporation
Mr. Bob Kelly, Senior Corporate Applications Engineer at Maxim Corporation
Mr. Daniel Kong, Strategic Customer Applications Engineer at Maxim Corporation

This project is supported in part by a contract with Maxim Corporation.

Submitted on June 7, 2001 at 5:00 PM

Executive Summary – We analyze and improve upon a Wideband Code-Division Multiple Access transceiver, dividing the task at the block level into three portions: receive, transmit, and frequency synthesis. Each portion is attacked by a team of two EE senior design students at UC Riverside, working with an industry engineer from Maxim Integrated Products. After performing preliminary research and analysis to become familiar with the reference architecture and design issues involved, each team works on a particular aspect of the design as described herein.

The design problem facing the receiver team is to improve adjacent interferers rejection at the I/Q baseband output of the receiver. To accomplish this, an active fourth-order low-pass Butterworth filter is designed, simulated, implemented, and tested. Of the numerous filter types available, this one is chosen for its gain, high linearity, maximally flat response in the pass-band, and minimal number of components required. The final design uses a Sallen-Key topology and provides 6 dB of gain in the pass-band, with a corner frequency of 3 MHz. A spreadsheet is created for performing the necessary calculations, and *PSpice* is used for simulation of the design. The circuit is realized in hardware, its characteristics extracted with an RF network analyzer, and fine-tuned to meet specifications. The realized circuit obtains 39 dB of attenuation at 15 MHz, meeting the most stringent specifications for the design.

The goal of the transmitter team is to design an IF filter in the transmit path of the radio. The filter must have lower cost and improved performance over the currently used SAW filter, in rejecting harmonics and other spurious signals generated by the quadrature modulation process. To meet the goal, analysis is performed to determine filter specifications and a passive second-order narrowband band-pass Butterworth filter is designed, simulated, and verified in hardware. A capacitive-coupled resonator architecture is chosen for its design characteristics and ease of tuning. A spreadsheet is created to automate design calculations and *PSpice* is used for simulation. Because of the high frequency range involved, special care is taken to account for finite inductor Q and stray capacitance present in the physical circuit. The prototype obtains a 3 dB bandwidth of 35.4 MHz, and 40 to 44 dB of attenuation at 190 and 760 MHz, respectively.

The design problem facing the frequency synthesizer team is to provide four clean local oscillators for the down- and up-conversion of the RF and IF signals in both the receive and transmit paths. To accomplish this goal, phase locked loops are used to produce clean signals at much higher frequencies than obtainable from crystal oscillators. An Integer-N PLL architecture is chosen for its low power consumption, small physical size, and economy. A third-order lead-lag passive low-pass filter is designed because the loop filter is key to generating an output with minimum noise. Key criteria include the amount of phase noise generated, the amount of spurious noise suppressed, and the time taken to lock to a specified frequency. *Matlab* code is written and used to calculate key values and to run simulations for given system specifications. The final loop filter design for the transmitter IFLO features a 1.5 kHz bandwidth, netting –58 dBc/Hz phase noise attenuation and spurious noise at a 1 kHz offset with a settling time of 5 ms.

Keywords – Wireless communications, Wideband Code-Division Multiple Access, Transceiver architecture, Low-pass filter design, Band-pass filter design, PLL design.

Table of Contents

Page

Chapter One – Introduction *(Karl Doering)*
   1.1. Introduction 1
   1.2. A Historical Perspective 1
   1.3. A Glossary of Acronyms and Abbreviations 3

Chapter Two – Design and Technical Results
   2.1. Introduction 4
   2.2. Problem Statement 4
      2.2.1. Receiver *(Steven Gyford)* 4
      2.2.2. Transmitter *(Michael Wilson)* 4
      2.2.3. Frequency Synthesizer *(Oscar Servin)* 5
   2.3. Design Specifications 5
      2.3.1. Receiver *(Steven Gyford)* 5
      2.3.2. Transmitter *(Michael Wilson)* 7
      2.3.3. Frequency Synthesizer *(Heath Deuel)* 8
         2.3.3.1. Radio Frequency Local Oscillator (RFLO) 8
         2.3.3.2. Receive Path – Intermediate Frequency Local Oscillator (Rx-IFLO) 8
         2.3.3.3. Transmit Path – Intermediate Frequency Local Oscillator (Tx-IFLO) 8

Chapter Three – Method of Solution
   3.1. Introduction 10
   3.2. Our Design 10
      3.2.1. Receiver *(Nadim Addus)* 10
      3.2.2. Transmitter *(Karl Doering)* 11
      3.2.3. Frequency Synthesizer *(Oscar Servin)* 13
         3.2.3.1. Integer-N Phase Locked Loop 14
         3.2.3.2. Crystal Oscillator 15
         3.2.3.3. Phase Frequency Detector 15
         3.2.3.4. Loop Filter 16
         3.2.3.5. Voltage Controlled Oscillator 16
   3.3. Alternative Approach / Design Trade Off 17
      3.3.1. Receiver *(Steven Gyford)* 17
      3.3.2. Transmitter *(Karl Doering)* 18
      3.3.3. Frequency Synthesizer *(Oscar Servin)* 19
   3.4. Lessons Learned 20
      3.4.1. Receiver *(Nadim Addus)* 20
      3.4.2. Transmitter *(Karl Doering)* 20
      3.4.3. Frequency Synthesizer *(Oscar Servin)* 20

Table of Contents (continued)

Chapter Four – Evaluation
- 4.1. Introduction 21
- 4.2. Discussion of Results / Test Plan 21
    - 4.2.1. Receiver *(Nadim Addus)* 21
    - 4.2.2. Transmitter *(Karl Doering)* 23
        - 4.2.2.1. Initial Plan 23
        - 4.2.2.2. Intermediate Trials 23
        - 4.2.2.3. Final Plan 25
    - 4.2.3. Frequency Synthesizer *(Heath Deuel and Oscar Servin)* 26
        - 4.2.3.1. Test Plan 26
        - 4.2.3.2. Discussion of Results 26
- 4.3. Design Comparison / Design Trade Off 29
    - 4.3.1. Receiver *(Nadim Addus)* 29
    - 4.3.2. Transmitter *(Karl Doering)* 33
    - 4.3.3. Frequency Synthesizer *(Heath Deuel)* 35

Chapter Five – Administrative *(Michael Wilson)*
- 5.1. Introduction 37
- 5.2. Cost Analysis 37
- 5.3. Organization of the Task 38

Chapter Six – Meeting Expectations
- 6.1. Introduction 39
- 6.2. Design Constraints 39
    - 6.2.1. Receiver *(Steven Gyford)* 39
    - 6.2.2. Transmitter *(Michael Wilson)* 40
    - 6.2.3. Frequency Synthesizer *(Heath Deuel)* 41
- 6.3. Elements of Design *(Steven Gyford)* 43

Chapter Seven – Conclusions
- 7.1. Introduction 44
- 7.2. Expansion and Improvement 44
    - 7.2.1. Receiver *(Nadim Addus)* 44
    - 7.2.2. Transmitter *(Michael Wilson)* 44
    - 7.2.3. Frequency Synthesizer *(Heath Deuel and Oscar Servin)* 44
- 7.3. User's Manual 45

Table of Contents (continued)

Appendix A: Parts List 46
Appendix B: Equipment List 47
Appendix C: Software List 47
Appendix D: Special Resources 47
Appendix E: Parts Cost Breakdown 48
Appendix F: *Matlab* Code for Loop Filter Design 49

References 51

Acknowledgement 53

Chapter One – Introduction

1.1. Introduction

This project involves the analysis of and improvement upon a Wideband Code-Division Multiple Access transceiver. We divide the tasks among ourselves at the block level into three portions: receive, transmit, and frequency synthesis. Each portion is attacked by a team of two EE senior design students at UC Riverside working with an industry engineer from Maxim Integrated Products. After performing preliminary research and analysis to become familiar with the reference architecture and design issues involved, each team works on a particular aspect of the design as described in this paper.

In the remainder of this chapter we give a brief historical perspective followed by a glossary of acronyms and abbreviations we frequently use. Chapter Two introduces the specific design problem of each team, providing a general problem statement followed by precise design specifications. In Chapter Three the method of solution for each team is presented, beginning with an overview of the chosen design solution among a variety of possible solutions. The designs are presented in detail, with the key features and functions of each component emphasized. In Section 3.3 alternative approaches to the design problem and the tradeoffs involved are examined more closely. Chapter Three closes with insight into some of the valuable lessons learned in the design process.

In Chapter Four, each team conducts a thorough evaluation of its design. Results from both *PSpice* simulations and hardware measurements are presented, compared, and contrasted. Both strengths and weaknesses in the design are stated and evaluated with respect to alternative methods of solution available.

Chapter Five deals with key administrative issues, including a complete cost analysis and a description of task organization and scheduling. In Chapter Six, the design constraints are reexamined. Discrepancies in the results from the original design specifications are examined and accounted for, and the corresponding effect on the final outcome is analyzed. The chapter closes with a table addressing elements of design, including economic factors, safety, reliability, aesthetics, ethics, and social impacts.

We conclude in Chapter Seven by discussing the impact of our work and possible future expansions.

1.2. A Historical Perspective

Since 1901, when Marconi first transmitted radio signals across the Atlantic Ocean, radio communication has progressed dramatically [14]. Early radio architectures were very simple and could be easily built out of one's garage with a few components. As radio communication became more popular and more and more communication by radio took place, people soon realized that the electromagnetic spectrum is a precious resource which must be shared among all involved. Frequency plans were created, and higher and higher frequency ranges were explored.

Design techniques advanced, and new architectures were developed to handle the problems encountered with higher-frequency designs.

Early radios worked in the kHz range, but today's W-CDMA architecture operates in the GHz range. Simple LC filters cannot be built to select narrow channels at such high frequencies, and techniques such as down-conversion, up-conversion, and intermediate frequencies are required, leading to the superheterodyne architecture commonly used today.

1.3. A Glossary of Acronyms and Abbreviations

3GPP	Third Generation Partnership Project
ACS	adjacent channel selectivity
Atten	added spurious attenuation due to R_3 and C_3
CDMA	code-division multiple access
dBc	decibels relative to carrier
dBm	decibels referenced to one milliwatt
DSP	digital signal processor
F_{comp}	comparison frequency
FDD	frequency division duplex
F_{out}	output frequency of the VCO
F_{ref}	crystal reference frequency
I/Q	in-phase and quadrature
IC	integrated circuit
I_{cp}	phase detector / charge pump gain
IF	intermediate frequency
IFLO	intermediate frequency local oscillator
IP_3	third-order intercept
K_{VCO}	VCO gain (tuning sensitivity)
LNA	low-noise amplifier
LO	local oscillator
Mcps	million chips per second
N	N divider value
PCB	printed circuit board
PFD	phase frequency detector
PLL	phase locked loop
Q	quality factor
R	R divider value
RF	radio frequency
RFLO	radio frequency local oscillator
RMS	root mean square (reference to phase noise)
Rx	receive
SAW	surface acoustic wave (filter)
SMT	surface-mount technology
Tx	transmit
TXCO	temperature-compensated crystal oscillator
VCO	voltage-controlled oscillator
w_c	true closed-loop bandwidth (rad/s)
W-CDMA	wideband code-division multiple access

Chapter Two – Design and Technical Results

2.1. Introduction

Over the course of six months, the Maxim / UCR (Maxim 6) project has studied a transceiver on a range of issues. These issues have been examined from the transmit, receive and control perspectives, based upon a reference design of a W-CDMA quadrature superheterodyne transceiver.

This chapter contains the problem statement and the presentation of the desired system specifications. The purpose of Section 2.2 is to discuss what our problem entails and give the reader a general understanding of what we are trying to solve. Section 2.3 illustrates the specifications for each component.

2.2. Problem Statement

Here we describe the technical problem to be solved for each of the three components of the transceiver, namely the receiver, the transmitter, and the frequency synthesizer components.

2.2.1. Receiver

The receiver portion of the overall project requires us to verify that the receiver in the W-CDMA superheterodyne transceiver reference design meets the Third Generation Partnership Project (3GPP) technical specifications through analysis of the system components. For this analysis we include factors influencing the architecture such as intermodulation, linearity, noise figure, and gain. Calculations are done to determine the effect of these factors on the cascaded system, as limitations of system performance due to these factors require the optimization of the design process to achieve better RF performance.

Once the receiver architecture has been validated, the second goal of this part of the project is to design a low-pass filter to be implemented at the output of the quadrature demodulator in the baseband I/Q signal path to further improve adjacent channel interferers rejection. Requirements for defining the filter characteristics are derived from the 3GPP specifications. These requirements determine the design of the filter, which are verified through simulation first, and secondly through measurements of a physical circuit's performance.

2.2.2. Transmitter

The efforts of the transmitter portion of the W-CDMA transceiver project are divided into two stages. The first stage is a comprehensive signal path analysis of the transmitter portion of the Maxim W-CDMA reference design in [9]. The signal path analysis enables us to master many of the fundamental concepts required in the design of a state-of-the-art radio transceiver system. The second stage of the project is to use these concepts of radio frequency design in combination with advanced techniques of high frequency filter design to design and implement a band-pass filter to replace the SAW / BPF in the IF portion of the transmitter.

2.2.3. Frequency Synthesizer

The reference design for the transceiver contains three PLLs to produce the four local oscillator signals that are needed. The primary result for the frequency synthesizer portion is to provide a clean signal for the transmit and receive path for up- and down-conversions of RF to IF signals [24]. We have designed the loop filter for all three PLLs and were lucky enough to test one. We have demonstrated that a fairly clean signal, which maintains a range of 30 kHz, can be generated by our VCO by implementing a good loop filter. An initial prototype of the TxIF / LO loop design has demonstrated that a signal of 760 MHz, with –58.83 dBc phase noise and negligible spurious noise can be produced to drive our modulator while the RF / LO up-converts the modulated carrier to the on-channel RF frequency for antenna transmission.

The goals for this project are threefold.

We first need to determine the main criteria behind each component of the PLL / LO. In order to meet our goals, it is necessary to understand and select the correct component of our PLL. When selecting the correct components we need to take into consideration the application of our design and specification of our system.

Understanding the frequency plan and justifying its effects on the radio is the second thing that is accomplished. The main focus of our design revolves around this idea. Understanding how and why the frequency must be down- and up-converted in certain areas is inevitable for the correct functionality of our radio [14].

Lastly, we need to calculate the control loop parameters given some specifications. The way that we will physically get a clean local oscillator to be mixed with our RF to provide the proper IF signal is by not only selecting the proper PLL components, but also having a good filter design. Selection of good components will help to avoid phase noise and the design of a good loop filter will allow us to eliminate spurious noise [1]. Further discussion on noise is to follow in the next chapter.

2.3. Design Specifications

Here we provide precise specifications for each of the components to be designed.

2.3.1. Receiver

The specifications that describe the attenuation characteristics for the low-pass filter, which show us the channels, their frequency offsets and the attenuation required at those particular frequency offsets, are given in Table 2.3.1-1. These characteristics are taken from the 3GPP technical specifications.

Table 2.3.1-1. Frequency and attenuation characteristics.

Channels	Freq Offset (MHz)	Att. LPF (dBc)
CH-Wanted	0	0.0
CH-Adjacent	± 5	-10.0
CH-In-Band-Blk	± 10	-25.0
CH-In-Band-Blk	± 15	-45.0
CW-Out-Band1-Blk	>± 15	-45.0
CW-Out-Band2-Blk	>± 60	-55.0
CW-Out-Band3-Blk	>± 85	-55.0

In addition to these specifications, we must also decide on a filter type, implementation, topology, and order. Another specification requires the filter to have a single-ended output as well as a voltage gain of 2 V in the pass-band. Since we are filtering both the in-phase and quadrature signals, two filters are needed. However, in order for the signals to be processed properly, they are required to be no more than 5 degrees out of phase and be within 0.5 dB of each other in magnitude. The final criterion to consider is the type of filter to be used for our design.

The first specifications that the 3GPP paper describes are the frequency bands and channel arrangement. The chip rate is set to 3.84 Mcps for all the information provided in the specifications [12]. For the receiver portion of the transceiver, the frequency band is set to 2110 to 2170 MHz with a separation of 190 MHz between the transmit and receive bands. Channel spacing is set to 5 MHz with a channel raster of 200 kHz [25].

The next set of specifications that are described are the requirements for the receiver. These are the specifications that will be used to determine whether or not the entire receiver chain is valid. The dedicated physical channel power required is $P_{DPHC\_Ec} = -117$ dBm / 3.84 MHz [25]. This is the minimum power for a signal that this receiver is required to be able to detect. In addition to the minimum levels, the maximum level is also defined to be -25 dBm / 3.84 MHz [25]. From knowing the minimum power level required by the receiver, we can determine that the noise figure (NF) for the entire receiver is 9 dB [12]. With a loss of 4 dB in the duplexer, the NF then becomes 5 dB for the rest of the receiver chain [12].

Adjacent channel selectivity (ACS) is another important factor to consider in that it determines whether or not the receiver can or cannot receive a desired signal in the presence of an adjacent channel signal. The specification for the ACS is used to determine the selectivity of the receiver at a frequency offset of 5 MHz. With a desired signal power of $P_{DPHC\_Ec} = -103$ dBm / 3.84 MHz and an adjacent signal power $P_{AC1} = -52$ dBm / 3.84 MHz, we can determine the selectivity of the receiver [12]. First the acceptable interference power is found, then the selectivity is found by subtracting the acceptable interference power from P_{AC1}. This yields a selectivity of 33 dB at ±5 MHz [12].

Another requirement of the receiver is the ability to receive signals in the presence of an unwanted interferer. These unwanted interferers are commonly called blockers. There are two types of blockers: in-band and out-of-band blockers. The minimum requirements for these blockers are described in the 3GPP specifications. The in-band blockers are centered at frequency offsets of ±10 MHz and ±15 MHz. Signal powers for the in-band blockers are –56 dBm / 3.84 MHz and –44 dBm / 3.84 MHz respectively [25]. Out-of-band blockers are located at frequency offsets of ±15 MHz, ±60 MHz, and ±85 MHz. Signal powers for the out-of-band blockers are –44 dBm / 3.84 MHz, –30 dBm / 3.84 MHz, and –15 dBm / 3.84 MHz respectively [25]. These blocking requirements can be used to determine the selectivity of the receiver at the frequency offsets described above.

Another specification that needs to be taken into consideration is the receiver's ability to receive the desired signal in the presence of third-order mixing products. Such specifications are known as intermodulation characteristics. The minimum requirements are defined for frequency offsets of ±10 MHz and ±20 MHz. The power of the signals at these frequencies is –46 dBm / 3.84 MHz [25]. From these specifications the selectivity of the receiver is determined to be 58 dB at 10 MHz and 20 MHz [12].

2.3.2. Transmitter

The signal path analysis performed on the transmitter and the design of the BPF for the IF portion of the transmitter are each accomplished with the consideration of certain required specifications. The signal path analysis is performed with the requirement for the transmitter to provide a signal output at a frequency ranging between 1920 MHz and 1980 MHz for a local oscillator up-convert frequency input range between 2300 MHz and 2360 MHz, respectively. Additionally, the transmitter output is required to maintain a minimum output level of +23 dBm with a minimum output level range of 75 dB over the entire operating frequency range [9].

The design specifications for the IF / BPF are derived from the signal path analysis performed on the transmitter. The IF / BPF prototype is used to replace the IF / SAW / BPF used in the reference design illustrated in [9]. To meet the design requirements, calculations of all spurious frequencies resulting from intermodulation and up-convert mixing functions are identified. The specifications of the filter design require the filter to have a center frequency of 380 MHz, an insertion loss of 8 to 12 dB, a 3 dB bandwidth of 6.8 MHz, and a minimum attenuation level of 30 dB at 360 MHz and 400 MHz. The design specifications are summarized in Table 2.3.2-1.

Table 2.3.2-1. Transmitter design specifications.

Signal Path Analysis Specifications	
Operating Frequency Range	1920 MHz - 1980 MHz
Output Level	23 dBm
Minimum Output Level Range	75 dB

Filter Design Specifications	
Center Frequency	380 MHz
3 dB Bandwidth (minimum)	6.8 MHz
Attentation at 190 MHz	37 dB
Attenuation at 760 MHz	40 dB
Differential Input Impedance	400 Ohms
Differential Output Impedance	600 Ohms

2.3.3. Frequency Synthesizer

Three PLLs are used in the reference design of a superheterodyne transceiver that we use in our design. Each PLL uses the system clock crystal oscillator at 19.68 MHz to generate a LO signal displaying unique characteristics. Each LO signal must contain an –80 dBc/Hz phase noise attenuation.

2.3.3.1. Radio Frequency Local Oscillator (RFLO)

This LO must tune over a range of 2300 MHz – 2360 MHz while having a switching speed less than 1 ms. This LO is used for both transmit and receive paths. The comparison frequency for this PLL is restricted to the channel spacing of the incoming RF signal, which is 200 kHz.

2.3.3.2. Receive Path – Intermediate Frequency Local Oscillator (Rx-IFLO)

This LO is fixed and therefore has no tuning range. The output frequency is 380 MHz. The switching speed is not an essential element in the performance of this PLL and therefore the design constraint can be relaxed to 5 ms. The comparison frequency for this PLL is determined by the designer.

2.3.3.3. Transmit Path – Intermediate Frequency Local Oscillator (Tx-IFLO)

This LO is also fixed, at 760 MHz, with a switching speed of 5 ms. The comparison frequency for this PLL is determined by the designer.

Table 2.3.3.3-1. System specifications.

Local Oscillator	V_{out} (MHz)	Switching Speed	Phase Noise Attenuation
RFLO	2300 – 2360	1 ms	−80 dBc/Hz
Rx – IFLO	190	5 ms	−80 dBc/Hz
Tx – IFLO	760	5 ms	−80 dBc/Hz

Chapter Three – Method of Solution

3.1. Introduction

This chapter presents our designs as well as reasons to why we made the engineering choices that we encountered. Section 3.2 details the architectures of our designs and advantages and disadvantages of the chosen architectures. Components used in the design are presented along with reasons to why these were selected. In Section 3.3 we continue our discussion of other methods to approach the solutions and explanations as to why these were not picked for our design. Section 3.4 states the lessons learned in selecting components for our architectures.

3.2. Our Design

In this section each of the three teams presents its design, emphasizing the key features and development.

3.2.1. Receiver

The implementation of the low-pass filter is a key factor that must be taken into consideration. A fourth-order Butterworth filter is selected for our design. This implementation is selected for its maximally flat magnitude response in the pass-band. Once the filter type is determined, the filter topology is selected. The next specification we take into consideration is the type of filter that is used. We choose to use an active filter for its linearity and lack of inductor in the circuit. This simplifies our analysis and the overall approach to our design. In the next step, we use one of the reference books [18] for filter design by which we take the following approach:

1. Determining requirements.
2. Finding filter or design that meet these specifications.
3. Normalizing.

Filter tables are used to simplify our circuit design, based upon the idea of constructing a fourth-order Butterworth filter by cascading two second-order filters. The tables contain scaling factors for the corner frequencies and the required Q for each stage of our filter. After completion of these tasks we are able to calculate the circuit component values required. Our requirements for the filter were already given in Chapter Two of this report. So given those requirements, our next step is to calculate the order (n) of the filter using Equation 3.2.1-1.

$$n = \frac{\log_{10}\left(\frac{\sqrt{10^{\left(\frac{A(dB)}{10}\right)} - 1}}{\sqrt{10^{\left(\frac{A_c(dB)}{10}\right)} - 1}}\right)}{\log_{10}\left(\frac{F_s}{F_c}\right)} \qquad (3.2.1\text{-}1)$$

Factors affecting this filter order include: cutoff frequency F_c, attenuation at cutoff A_c (dB) [4], the frequency F_s, and attenuation at stop-band A_s (dB). All these specifications must be met. After calculating filter order, the filter element values are looked up in a filter table. The topologies are given and we are able to come up with the circuit as shown in Fig. 3.2.1-1.

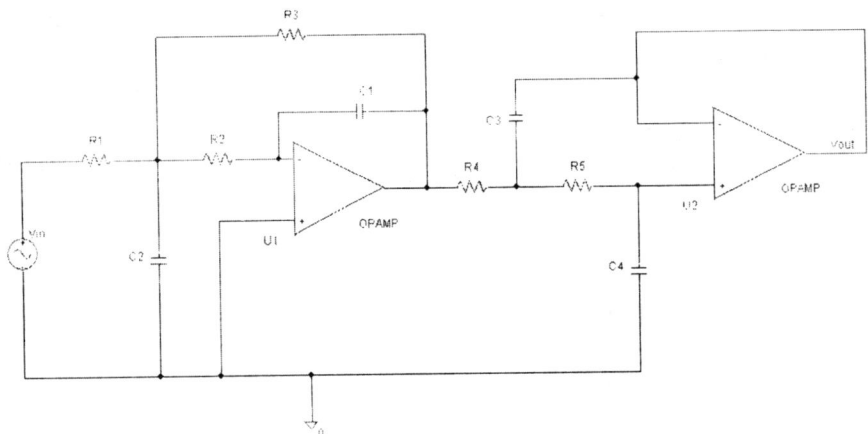

Fig. 3.2.1-1. Circuit diagram of the low-pass filter.

3.2.2. Transmitter

Our design consists of a passive band-pass IF filter in the transmit path as a replacement to the standard SAW filter originally present. Let us first examine the purpose of this filter.

From the W-CDMA reference design [9], this filter functions to suppress harmonics of the IF signal, resulting from the quadrature modulation process as well as any nonlinearities in the IF variable gain amplifier (VGA). Assuming the baseband I and Q channels are band-limited to, say, 5 MHz, at an IF of 380 MHz, undesired mixing products are expected to fall near DC and around 760 MHz and up. Therefore, the sharp response of the IF / SAW filter may not be necessary. For reasons of cost, replacing the SAW filter with a low-order LC filter is an attractive option.

In fact, at the frequencies at which harmonics are expected, a simple LC filter may even surpass the performance of a SAW filter. SAW filters tend to have rather high insertion losses and relatively flat response outside the pass-band. However, with an LC band-pass filter, attenuation generally increases the farther away from the center frequency.

Given the high center frequency at which this filter must operate, designing a practical passive narrowband filter is quite a challenge. Component values (and in particular inductor Q) are often either very sensitive or unrealizable, or both. There are a wide variety of filter types and design techniques available, including approximations, normalizations, and transformations as given in [7]. After much reading and experimentation with circuit simulators, the capacitive coupled resonator configuration was chosen. Such an architecture is desirable given our specifications, as it generally yields more practical values than low-pass to band-pass transformations and tuning is simplified since all nodes are resonated to the same frequency [18].

To determine the necessary order of the filter, we make use of an Excel spreadsheet, which we designed specifically for this purpose. This eliminates the tedious calculations required as well as the possibility of error in calculation, and makes it very easy to "feel out" the design by instantly observing the effects of modifying key parameters.

Our design process is as follows. The upper and lower pass-band frequencies f_u and f_l (respectively) are entered into the spreadsheet. The center frequency $f_0 = \sqrt{f_l f_u}$, pass-band bandwidth $BW_{3dB} = f_u - f_l$, and pass-band Q, $Q_{bp} = f_0 / BW_{3dB}$ are then filled in. Before continuing further, we examine Q_{bp} to determine whether the final circuit will be realizable. (If not, then either the specifications must be relaxed or a different implementation must be used.)

Next, the type and order of the filter must be determined. This task is accomplished based upon the band-pass steepness factor A_s = (stop-band bandwidth) / BW_{3dB}, the complexity (order) required to meet the specifications, and by examining reference design tables in [18] to choose a filter type based upon the desired phase and magnitude response characteristics.

Based upon our specifications and the response characteristics available, we initially examine a fourth order 0.1 dB Chebyshev resonator, a third-order traditional Gaussian to 6 dB resonator, and a second-order Butterworth resonator. Since we want a low-cost filter with flat phase response in the pass-band, high attenuation far from f_0, non-critical A_s, and a circuit simple enough to prototype with 0402 and 0603 surface-mount components, we choose the second-order Butterworth capacitive coupled configuration, Fig. 3.2.2-1.

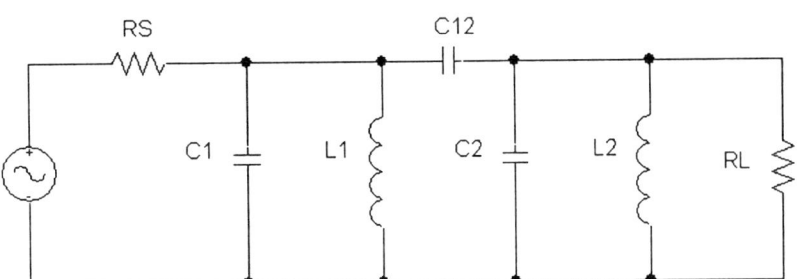

Fig. 3.2.2-1. Second-order base circuit configuration.

We obtain component values for this circuit using standard design tables and equations given in [18], and simulate the circuit in *PSpice*, producing the waveform in Fig. 3.2.2-2.

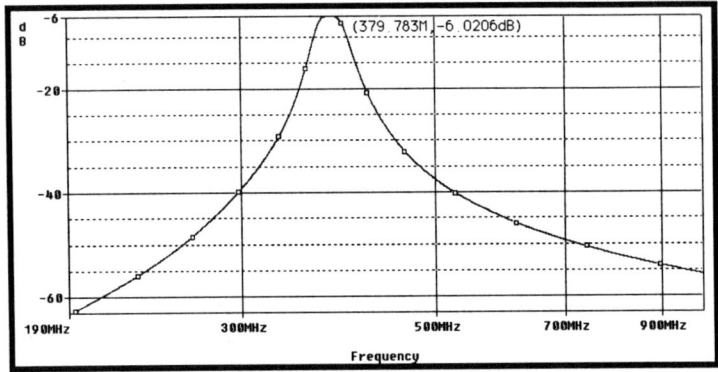

Fig. 3.2.2-2. Simulated response, 190 MHz – 1 GHz.

We then repeat the simulation a number of times, using standard component values at the limits of their specified tolerances, to obtain some indication of how sensitive the design is relative to component variations. Finally, we add additional capacitances and resistances to account for finite inductor Q, and stray capacitance present in a physical implementation, and we transform the circuit to its differential form. Our final circuit is shown in Fig. 3.2.2-3. (It is to be noted that the unlabelled components are not physically present in the circuit; these account for the effects just mentioned.)

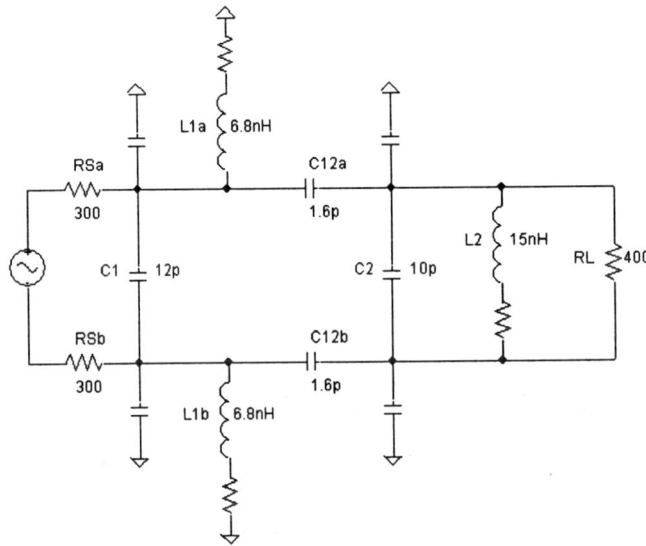

Fig. 3.2.2-3. Final circuit, after transformation to differential form.

3.2.3. Frequency Synthesizer

The best way to approach our problem is to use an Integer-N phase locked loop [22]. The components that make up our Integer-N / PLL are a crystal oscillator, phase-frequency detector, a VCO, low-pass loop filters, and dividers. The RF and IF / LOs will be similar componentwise, as well as in functionality.

A phase locked loop is a negative feedback loop, which tracks the error and corrects itself as changes are made to the input of the system. For phase locked loops, the quantity of interest is the phase of the input and output. The purpose of the phase locked loop (PLL) is to track the phase of the input and to try to match it, so that the output phase is the same [3]. An ideal PLL has no change in phases.

The simplest architecture of a PLL consists of three blocks. The first is the Phase Detector. The function of the Phase Detector (PD) is to detect the phase of the input. The PD tries to measure the phase error between input and output signals. The next block is a Loop Filter, normally a low-pass filter, but that depends on the type of PLL that is being built (more on different types of PLLs in Section 3.3.3). The purpose of the filter is to suppress the higher-order harmonics. The last block in our PLL architecture is the VCO. The Voltage Controlled Oscillator is used to control the phase of the output.

The Integer-N Phase Locked Loop is a basic feedback loop that can produce a low-noise local oscillator (LO) signal at a specific frequency, which can be mixed with an incoming Radio Frequency (RF) signal. A PLL's performance is characterized by how fast it locks to a new frequency and by how much noise is exhibited by the output, also known as phase noise [3]. The Integer-N / PLL gets its name from the output frequency being an integer multiple of the reference input frequency.

3.2.3.1. Integer-N Phase Locked Loop

The Integer-N Phase Locked Loop is composed of four basic functional blocks: a phase frequency detector (PFD), a loop filter, a voltage-controlled oscillator (VCO), and a programmable divider (N). A general block diagram of an Integer-N / PLL is shown here in Fig. 3.2.3.1-1.

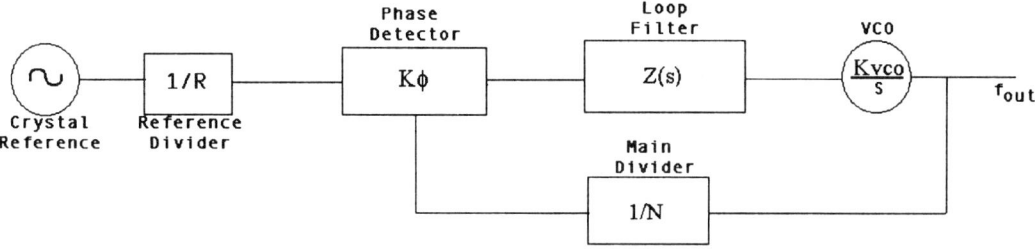

Fig. 3.2.3.1-1. Integer-N / PLL block diagram.

One of the most intriguing capabilities of the PLL is its ability to suppress noise superimposed on its input signal [1]. The PLL is a closed-loop feedback system that uses a Phase / Frequency Detector (PFD) to compare and adjust the system.

The Integer-N / PLL architecture is desired over other types of architecture for its advantages. The Integer-N / PLL provides functionality, low power consumption, savings in space, and it's economic [1]. Therefore, it has become the basis of wireless communication since it operates at low voltages and low currents. The switching speeds can be enhanced to the sub-millisecond range. This architecture does have its disadvantages however. One major disadvantage is that the phase noise performance degrades significantly, because of its high division ratio.

3.2.3.2. Crystal Oscillator

In order for the PLL to track, we are required to have a good reference signal to be able to get a good lock on our system. We have found that crystal oscillators, typically flat disc-shaped quartz crystal resonators, provide a very clean sinusoidal wave. We have found that the Q of such an oscillator is typically greater than 2000, which means that the frequency stability is excellent and the phase noise at 1 kHz offset is superb [14]. The crystal oscillator provides excitations and resonance modes which are predictable, repeatable and most importantly, exploitable. For our design, we are required to have a crystal oscillator at 19.68 MHz. We cannot get any higher oscillation since this serves as the system clock for all digital components. The clock is also used in the microprocessor, which cannot read anything higher than 20 MHz. The drawback and reason why we do not use a crystal oscillator in place of our PLL is that the crystal tuned oscillator exhibits a natural resonant frequency well below 100 MHz, which is undesired for our design.

3.2.3.3. Phase Frequency Detector

The Phase Frequency Detector (PFD) is mostly used in PLL applications for several reasons. PFDs offer unlimited pull-in range, which guarantees PLL acquisition under even the worst operating conditions [3]. Secondly, because the output signal of the PFD depends on phase errors in the locked state of the PLL and on frequency errors in the unlocked state, a PLL which uses PFD will lock under any condition, irrespective of the type of loop filter used.

In order to understand how a PFD works, we must take a look at it in a more detailed manner. A PFD is composed of two D flip-flops, which take in four inputs. One input of each D flip-flop is set to the logical high, and the other input from each D flip-flop comes from the outputs of our PFD. One output of each D flip-flop is connected to a logical AND gate, which in turn is connected to the reset pin of both D flip-flops. Then the output of one of the D flip-flops is connected to an NPN / MOS transistor, and the second D flip-flop output is connected to a PNP MOS transistor. When the D flip-flop whose signal is connected to the N-channel MOS transistor is set high, the transistor conducts, so the final output is at ground potential. When the D flip-flop whose output is connected to the P-channel MOS transistor is high, the transistor conducts and the final output is some positive supply voltage.

The PFD that we are using in our design has two inputs and two outputs. The two inputs are the reference frequency that comes from the crystal oscillator, and the output frequency from the Voltage Controlled Oscillator (VCO), which is the frequency being fed back in our negative feedback control loop (PLL). The outputs of the PFD are used to drive two switched current sources, each of which is known as a "charge pump" [14]. The charge pump places a charge on a capacitor by sinking or sourcing current according to the difference in phase measured by the PFD. The charge on this capacitor is used as the controlling voltage to the VCO. Additional capacitive and resistive networks may be added to ensure stability and suppress ripples.

3.2.3.4. Loop Filter

The loop filter is a passive second-order filter that outputs a DC voltage to the VCO. The PLL must rely on the loop filter to maintain a low phase noise and low spurious output [6]. The design of the filter also determines the transient response or switching time. One can increase the order of the filter to obtain better phase noise attenuation.

Since this is a passive filter it only requires a number of capacitors and resistors. For a second-order filter there are two capacitors and one resistor.

The values for the components in the loop filter can be calculated using different methods. Following National Semiconductor's application note 1001 [8], there are parameters that need to be determined before the actual filter components are calculated. One must know the Voltage Controlled Oscillator (VCO) Tuning Voltage constant, which is the frequency versus tuning ratio [26]. We also need to know the Phase Frequency Detector / charge pump current (K_ϕ), which is the ratio of the current output to the input phase differential [8]. The radio frequency output of the VCO at which the loop filter is optimized (RF_{out}) and the frequency of the phase detector inputs, which are usually equivalent to the RF channel spacing, are also needed to determine the loop filter components. Lastly, the final parameter that must be known is the main divider ratio, which is equal to RF_{out} / F_{ref} [8]. Once these parameters are known, we can use the following *Matlab* code to calculate the loop filter components:

```
N = F_out/F_ref;

T_1 = (sec(ϕ_p π/180)-tan(ϕ_p π/180))/1.256e5;
T_3 = √((10^(ATTEN/20)-1)/(2πF_ref)^2);
T_2 = 1/(w_c^2(T_1+T_3));

C_1 = (T_1/T_2)(K_ϕ K_vco)/(w_c^2 N))(√(1+w_c^2 T_2^2)/((1+w_c^2 T_1^2)(1+w_c^2 T_3^2)))
C_2 = C_1(T_2/T_1 - 1)
C_3 = C_1/10

R_2 = T_2/C_2
R_3 = 0.1C_2R_2/C_3
```

Fig. 3.2.3.4-1. *Matlab* code to calculate loop filter components.

3.2.3.5. Voltage Controlled Oscillator

If the output frequency of an oscillator circuit can be varied by an input voltage, then the circuit is called a voltage controlled oscillator. A variable oscillator frequency is necessary in radio frequency applications so that exact channel selection can be achieved. It exhibits a tuning gain, K_{VCO}, which is also referred to as the sensitivity of the VCO and has units of frequency per voltage. The performance of the VCO is measured by looking at how much noise, or unwanted distortion, is generated and introduced into the system by the VCO [26]. By itself the VCO has a tendency to drift from the desired frequency and therefore must be controlled using modern control system techniques. The goal of the VCO is to produce a noise-free signal that can be used as a local oscillator that will aid in the down- or up-conversion of a received signal.

The basic hardware of an oscillator requires an LC circuit that feeds to the collector of an NPN transistor. Usually there is a feedback wire from the LC circuit to the emitter of the same transistor. To obtain a variable frequency output, a tank circuit is incorporated into the circuit. This tank circuit has a role similar to the flywheel in a car, in that it is required to start the oscillation. This circuit consists of an inductor and capacitor in parallel, which gets an input from a varactor or reversed-biased diode. The duty of the tank is to store and release a pure sinusoidal signal.

Three different VCOs are required in our transceiver. The output frequency of the VCO, which produces the RFLO, needs to have a range of 60 MHz. The input voltage necessary to achieve this range is 0.5 to 2.5 volts, resulting in a gain of roughly 30 MHz/V [27].

3.3. Alternative Approach / Design Trade Off

Most engineering designs have a number of approaches available, and every engineering design consists of a number of design decisions and tradeoffs. In general, RF designs contain tradeoffs between noise, power, linearity, frequency, voltage, and gain [14]. Now, we discuss specific alternatives and tradeoffs we encountered in our designs.

3.3.1. Receiver

There are many different implementations of filters as well as many different types of filters in the market. For this project, three different filter implementations were explored. While we chose to use a Butterworth filter, we also looked into the possibility of using a Chebyshev or a Bessel filter. The three different types of filters that were considered for this project were active, passive, and digital filters. As we have already discussed, we selected an active filter, and in the following we discuss our reasons for not choosing a passive or a digital filter type.

Here we discuss two other filter implementations and the various advantages and disadvantages of each one. Chebyshev is one of the filter implementations that could have been used for our low-pass filter design. While it has the advantage of better attenuation in the stopband, there is more ringing in the pass-band than that of the Butterworth implementation [17]. Ringing is undesired because it corrupts the signal, which is undesirable since our signal is already of a small magnitude. For this reason, we chose not to design a Chebyshev low-pass filter. The other filter implementation we looked at is the Bessel low-pass filter. While the Bessel filter implementation has excellent pass-band response, the attenuation in the stop-band is not so good as the Butterworth implementation [17]. This poor attenuation in the stop-band would result in the need for a higher-order filter, adding to the complexity of the circuit. In trying to design an appropriate filter, space is also a limiting factor and outweighs any advantages that the Bessel filter would offer in the pass-band.

The other factor we explored was the type of filter that could have been used. The advantage to using a passive filter is that there is no power supply required, which would benefit our need for low power consumption. In addition to no power supply, passive filters can accommodate large bandwidths and create little noise [10]. One disadvantage to passive filters is that they require more space than an active filter of similar order. Another problem with passive filters is that they are difficult and time-consuming to design [10]. Though we spent a considerable amount of time designing an active filter, the process was straightforward. The other type of filter that we explored was the digital filter. The advantages to digital filters are that they are not temperature or time sensitive. Also, digital filters are easily implemented on a general-purpose computer or workstation and are easily altered to fit a new design specification, which saves time in the design process [19]. One disadvantage of a digital filter is the comparatively high cost for simple tasks. Since our design task is very simple for a digital filter, we have no need for such an advanced filter. A digital filter needs a high processing speed when the bandwidth is large [21]. In our case where the bandwidth is 3 MHz, a digital filter would be a poor choice. The last disadvantage with a digital filter is that the hardware is detailed and the software required is complex [21]. For this reason, the design of a digital filter would require too much time, and it would not be realizable.

3.3.2. Transmitter

As mentioned in Section 3.2.2 and above in 3.3.1, in filter design there are a number of alternative methods and architectures available. However, our band-pass IF filter has vastly different design constraints than does the low-pass filter in the receive path. Here also, like in the receive path we do not use a digital filter, but an active configuration is not a viable option either. Active filters are feasible only up to a few MHz, whereas LC filters are practical into the hundreds of MHz.

There are two main types of passive band-pass filters: wideband and narrowband. A wideband band-pass filter is implemented by cascading individual low-pass and high-pass filters. Narrowband band-pass filters cannot be implemented as such, because of the increase in loss at the center frequency as the ratio of upper to lower cutoff frequencies increases. Because we require a ratio of much less than 2, we design a narrowband filter.

With these considerations in mind, there are still a wide variety of configurations and techniques to choose from. For example, the low-pass to band-pass transformation [16] is often used to design such a filter from standard low-pass design tables. Such a filter may or may not be an all-pole filter. Architectures include parallel tuned circuits, series tuned circuits, and synchronously tuned filters. For example, we considered using an elliptic filter, but it suffers high inductance spread so as to be impractical above 100 MHz [13]. Although certain transformations can be performed to reduce this effect, the resulting circuit requires roughly twice as many capacitors, meaning increased cost and difficulty in construction.

We choose a narrowband coupled resonator configuration for its desirable characteristics for high-Q filters and for its simplified tuning (since all nodes resonate to the same frequency). Of the possible narrowband coupled resonator configurations, we choose the capacitive-coupled configuration over the inductive-coupled configuration. The circuit with fewer inductors generally takes up less space and is more economical.

3.3.3. Frequency Synthesizer

Part of our design was to understand why we picked the components that make our PLL. Understanding the advantages and disadvantages of such choices is a crucial part of developing our design. When picking components or methods of solutions that are already on the shelves, we found many that we did not choose as explained in this section.

When selecting the type or architecture, we found that the Fractional-N / PLL was a possible solution. What makes the Fractional-N / PLL appealing for our design is that it improves phase noise by increasing the reference frequency, which in turn improves switching speeds as well as loop bandwidth [23], [3]. If we recall, the Integer-N / PLL's major disadvantage is that it has a high division ratio, thus causing significant degradation in phase noise performance. We weighed the options and decided to stay with our Integer-N / PLL architecture, because the major disadvantage of the Fractional-N / PLL architecture is that it adds complexity in circuitry and adds spurious signals [6]. We decided that the disadvantage of the Integer-N / PLL architecture could be downplayed by decreasing the bandwidth of the loop filter. When trying to deal with more circuitry, we have to keep in mind that we want the lowest possible power consumption from our circuit, which means a smaller circuit. The less circuitry, the smaller the cell phones or laptops we can provide for customers. Once our decision was made on the type of architecture, we moved to selecting the proper components of our PLL.

The type of oscillator was one of the first components we looked at. We found many types of oscillators such as sinusoidal oscillators, op-amp / RC oscillators, LC oscillators and crystal oscillators among others. The choice for the reference oscillator was a fairly simple one because the range of frequency that we are working with eliminates most of the other choices, because stable and accurate RC and LC oscillators are almost impossible to design at such high frequencies. Not mentioning that each of these components, like any electronic device, produces noise, while the crystal oscillator does not.

One of the most important components in our PLL is the phase detector (PD). While searching for the correct phase detector, we again found numerous choices. Some of these choices were the EXOR gate and the JK flip-flop phase detector. Each of these had its own advantages and disadvantages that we had to pay close attention to. The advantage of the EXOR gate is that it provided phase tracking that could be maintained when the phase error is confined to the range $-\pi/2 < E < \pi/2$, assuming symmetrical signals [3]. The disadvantage is that the phase detector becomes severely impaired if the signals are not symmetrical, causing a reduction in loop gain, which means a smaller lock range and pull out range. The pull out range is the frequency range that causes the closed-loop system to unlock. If we chose the EXOR gate as our PD, we would be adding an additional constraint to our system, which is undesirable. The next possible choice for the phase detector is the JK flip-flop. The advantage of the JK flip-flop over the EXOR gate is that the symmetry of the signals is irrelevant since the JK flip-flop depends on the rising edge of the signal. The disadvantage, however, is that if the loop filter doesn't contain an integrating term, the pull-in range stays severely limited [3]. An integrating term in the loop filter adds not only complexity to our circuit, but it leaves no room for future changes. We therefore decided that we do not want to be limited by other unnecessary constraints and therefore chose the PFD instead. We must recall that the PFD's major advantage is that it provides virtually unlimited pull-in range, which will cause any PLL that contains a PFD to lock regardless of the loop filter that is being used.

To select the type of loop filter was based on the type of PLL components that were chosen. We could have chosen an active low-pass filter versus our passive low-pass filter. The tradeoff is that we added less noise to our system when we chose a passive low-pass filter.

3.4. Lessons Learned

All of us have learned important lessons from our experience gained in working on our design problems. We share a few of those lessons below.

3.4.1. Receiver

The most significant lesson learned is in our approach to selecting the best solution to our problem. We have learned that a top-down method develops a design by initially proposing a solution in terms of block diagrams. Then each block diagram is broken down into a more detailed and manageable piece than otherwise. This is especially true at the beginning of the project when time is crucial. By following this process we were able to eliminate any unnecessary steps pertinent to the required task.

3.4.2. Transmitter

Designing a low-order band-pass filter initially sounded simple to us, but the more we worked on the design problem the more we learned about all of the design choices, tradeoffs, and issues involved. The target frequency range of our filter made the task particularly challenging. Computer simulation tools proved very useful, once we discovered how to simulate in *PSpice* circuits containing nodes with no direct DC path to ground. Even so, some designs that looked very good on paper or even in the simulator were either not realizable or would not have worked nearly as expected. Many of our preliminary trials required unrealistic component values, were extremely sensitive to component variation (for instance, there is no 0.1487 ± 0.00005 pF capacitor in the market), or did not account for finite inductor Q. Additionally, stray capacitance and even the inductance of small jumper wires used in a physical implementation had to be taken into account.

3.4.3. Frequency Synthesizer

In designing a system, we must be very precise in the selectivity of the components that make up our design. This will eliminate error in future steps. The selection of the proper components allows us to reach our goals in a smoother and more direct way. The process of selecting our components helps to avoid phase noise, reduce power consumption and eliminate the need for large circuits. We have learned that part of a design is not only understanding the concepts, but also dealing with non-ideal properties, which adds additional constrains in our design that we did not necessarily take into account when running the simulation.

Chapter Four – Evaluation

4.1. Introduction

Due to the requirements for highly specialized and expensive equipment, we were unable to construct our prototypes or take measurements at our local facilities. Special arrangements had to be made to take measurements in an adequately equipped environment. Maxim Integrated Products invited us to their laboratories in Sunnyvale, CA for two days to construct our prototypes and to conduct tests.

This chapter is concerned with evaluating the results of both simulation and actual measurements. In Section 4.2 we describe the ways in which we constructed and tested our prototypes, presenting our results and discussing their significance. In Section 4.3 we give comparisons between the final design and the original specifications, stating design tradeoffs encountered.

4.2. Discussion of Results / Test Plan

4.2.1. Receiver

The cascaded evaluation is carried for the entire receiver architecture. One of the parameters evaluated is noise figure of the entire system. This is one of the factors that determines system sensitivity. The actual mathematical calculation uses the Friis equation [4],

$$F_{total} = F_1 + \frac{F_2 - 1}{G_1} \quad (4.2.1\text{-}1)$$

where F and G are the noise factor and numerical gain of each stage in which

$$NF = 10 \log_{10} F \quad (4.2.1\text{-}2)$$

is the noise factor expressed in dB. The actual calculated values for the various parameters in Table 4.2.1-1 resulted in the overall noise figure of 5.0 dB. Using the IP_3 for each stage, the cumulative IP_3 of the cascade was also calculated. Also, IP_3, which has frequencies located close to the desired signal, was calculated to help determine the amount of channel interferers that would have to be attenuated.

Table 4.2.1-1. Cascade analysis for W-CDMA receiver.

Cascade Analysis for W-CDMA Receiver

Stage Name	Duplexer	LNA	RF Bandpass filter	Mixer	IF Band pass filter	IF Amplifier/ Demodulator
Stage#	1	2	3	4	5	6
A_V (dB)	-2	15	-2.5	12.0	-10	35
A_P (dB)	-2	15	-2.5	12.0	-10	35
Cumulative A_V (dB)	-2.00	13.00	10.50	22.50	12.50	
Stage NF (dB)	3.00	1.70	2.50	7.00	10.00	6.36
Cumulative NF(dB)	5.0	2.37	9.45	6.95	-8.64	-28.64
Stage IP3 (dBm)	100	4.2	100	6.8	1000 <- Vrms ->	0.7
Cumulutive IP3	-4.12 dBm	-6.12 dBm	9.30 dBm	6.80 dBm	73.01 dBm	700.00 mVrms

There are many possible solutions to our design. As explained in Chapter Three, the approach we have taken in all of our analysis of the low-pass Butterworth filter is to make use of the Sallen-Key normalized low-pass prototypes. The Fig. 3.2.1-1 shows a complete schematic diagram, and the actual element values are given in Table 4.2.1-2. Our filter has an overall 3 dB gain with a cutoff frequency at 3 MHz.

Table 4.2.1-2. List of components and values used in the filter design.

Component	*PSpice* Simulation	Original Circuit	Final Circuit
C_1	53 pF	68 pF	68 pF
C_2	0.186 nF	22 pF	22 pF
C_3	0.138 nF	8.2 pF	22 pF
C_4	20 pF	56 pF	68 pF
R_1	461 Ω	1 kΩ	1 kΩ
R_2	131 Ω	332 Ω	332 Ω
R_3	923 Ω	2 kΩ	2 kΩ
R_4	1 kΩ	2.7 kΩ	1 kΩ
R_5	1 kΩ	2.7 kΩ	1 kΩ
U_1	Ideal op-amp	THS4121	THS4121
U_2	Ideal op-amp	THS4121	THS4121

The filter meets all specifications listed in Table 2.3.1-1. All capacitors in the circuit, which are the main contributor of weight, were found to have values in the range of picofarads. This addresses some of our application considerations that deal with the issue of lower weight as well as smaller size. In summary the active design performed very well and gave us an overall cascaded performance that closely approximates the results obtained by calculation.

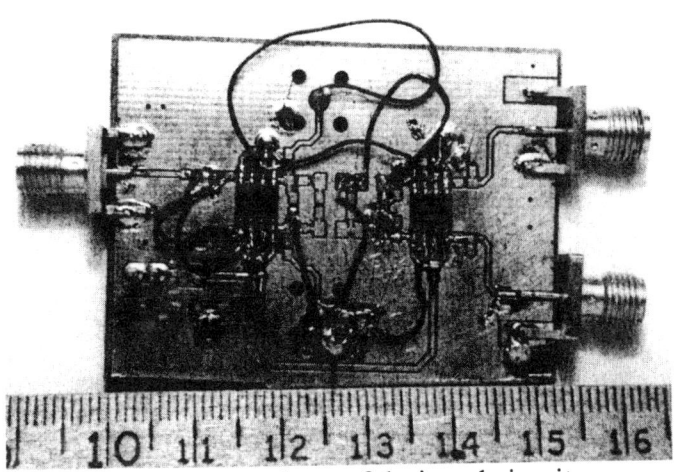

Fig. 4.2.1-1. Photo of designed circuit.

A test condition dictating the blocking performance required of the receiver was applied at 15 MHz. The Fig. 4.2.1-2 shows the actual performance under this test condition. When the receiver tries to process a weak signal in the presence of interferer, the weak desired signal tends to experience a vanishingly small gain. When the gain drops to zero, the signal is blocked [14].

So the receiver must be able to withstand a blocking signal of a certain level in dB. As shown in Fig. 4.2.1-2, the receiver meets our requirement. We are able to receive the wanted signal in the presence of an unwanted interferer at the specified frequency.

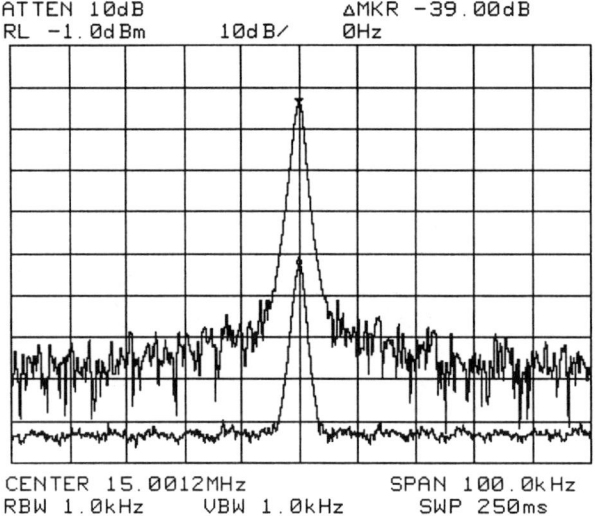

Fig. 4.2.1-2. Performance result obtained from network analyzer.

4.2.2. Transmitter

Shortly before heading up to the laboratories in Sunnyvale, we drove south to visit our advisor Mr. Bob Kelly in preparation. This consultation proved to be very beneficial and a definite test plan was formulated. However, upon arriving in Sunnyvale, we were given a MAX2360 evaluation kit to work with directly, and our test plan required substantial revision.

4.2.2.1. Initial Plan

The initial test plan was as follows: first, we construct a prototype of our filter on unetched FR4 circuit board. Physical construction consists of cutting grooves in the surface of the copper-plated board for electrically-open connections, which are verified with an ohmmeter. The surface-mount parts are then laid out and soldered in place. Measure the response of the filter by itself, then connect the filter to the MAX2360 evaluation kit and examine the effect of the prototype filter on the performance of the transmitter.

4.2.2.2. Intermediate Trials

Given the MAX2360 EV board, we attempted to implement the filter directly on the board by finding suitable traces and pads that can be reused for our purposes. This is a major constraint, but also one commonly encountered in practice, where a prefabricated board requires modification to function properly. Upon determining a physical layout given the constraints, we constructed our differential filter as given in Fig. 3.2.2-3.

Fig. 4.2.2.2-1. MAX2360 EV board.

Our modified board was connected per the standard EV test setup, and we attempted to lock the on-chip PLL to twice the IF, or 760 MHz. However, we could not lock to the desired frequency. In fact, the evaluation board used specifies an IF of up to only 300 MHz. For the PLL to lock at 780 MHz, the tank circuit requires modification.

After performing the necessary modification, the PLL locked at 760 MHz, but the range over which it locked was too narrow to be of use in determining the response of our filter at critical frequencies. Therefore, yet another modification was performed to bypass the on-board PLL circuitry, such that the signal can be injected directly from a high-quality signal generator. Thus the signal range was no longer limited by the on-board circuitry, but by the range of the Agilent E4433B signal generator.

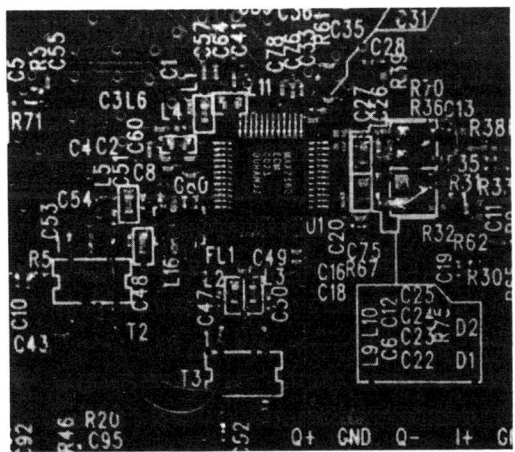

Fig. 4.2.2.2-2. Circuitry with all modifications.

Once all modifications were complete, in order to determine the response of our differential filter, we attempted to measure the magnitude of the input and output signals with an RF probe. In theory, the magnitude response was given by the difference in these two values. In practice, this proved to be very difficult, and we were unable to obtain consistent data. Furthermore, the undesired inductance resulting from the two long blue jumper wires at this frequency could have only made matters worse!

4.2.2.3. Final Plan

From our intermediate trials, evidently our initial plan to build and test the filter off-board first is most appropriate. Rather than constructing a layout from scratch, we use a prefabricated PCB with suitable pads and traces (courtesy of Mr. Daniel Kong), as illustrated in Fig. 4.2.2.3-2.

The filter circuit is reconstructed on the PCB using the design values, conversion is made to 50 Ω single-ended input and output, and SNA connectors are attached. The response is measured on a HP 8753E RF network analyzer. Due to the finite Q of real inductors and stray capacitance in the prototype, component values are tweaked as necessary until a suitable response is obtained. The final circuit is given in Fig. 4.2.2.3-1, where the unlabeled components are representative of these effects.

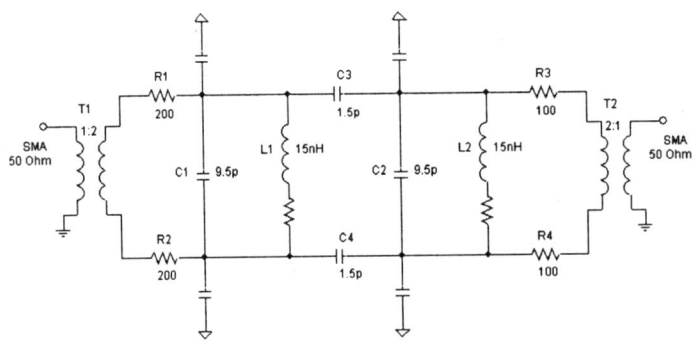

Fig. 4.2.2.3-1. Final prototype circuit.

Fig. 4.2.2.3-2. Implementation of prototype (dime shows scale).

4.2.3. Frequency Synthesizer

The setup for the lab limited the way we tested our prototype. In order to measure the required fields for our design, we are required to have the entire transceiver already built for us.

Understanding the test bench is as important as understanding the actual design. We shall explain the test bench so that we can reference it in future sections. To generate our reference signal, we have two possible choices for signal generators. The first is an E4433B Agilent ESG-D Series Signal Generator with a frequency range of 250 kHz – 4.0 GHz. The second is a Hewlett Packard 8648C Signal Generator with a frequency range of 100 kHz – 3200 MHz. To view the frequency response we used an 8562EC Agilent Spectrum Analyzer with a frequency range of 30 Hz – 13.2 GHz. Our test bench included a PC with control software.

4.2.3.1. Test Plan

The structure of our design requires us to have the entire transceiver built. In order to implement our filter design, we use the MAX2360 EV board. This is an evaluation board that has been laid out specifically for testing the MAX2360 quadrature transmitter chip by potential customers. This board is an acceptable piece to test our IFLO / PLL design. Even though we had designed and simulated the RFLO and both Rx and Tx-IFLOs, the MAX2360 EV board only allows us to test our Tx-IFLO. The layout of the board calls for size 0402 surface mount monolithic chip capacitors and resistors. This means that we need to round our design values according to the 5% standard production values, shown in Table 4.2.3.1-1. With our resistor and capacitor values chosen, we replace the existing loop filter components with our own values.

The next step in testing our design is the proper setup of the PC Control Software. This is just a graphical user interface program that allows us to use the processor of a PC in place of a microprocessor that would be present in a production unit. After the program was set to the proper settings we attach all the proper signals and analyzer fittings.

Table 4.2.3.1-1. 5% Values for resistors and capacitors.

5% Resistor Values:									
1	1.2	1.5	1.8	2	2.2	2.4	2.7	3	3.3	3.6
3.9	4.3	4.7	5.1	5.4	5.6	6.2	6.8	7.5	8.2	9.1

5% Capacitor Values:									
1	1.2	1.5	1.8	2.2	2.7	3.3	3.9	4.7	5.6	6.8	8.2

4.2.3.2. Discussion of Results

We discuss the Tx-IFLO, which was the only one that was measured and tested. The first step in getting our measurements was to get our PLL to lock to any frequency [1]. Fortunately for us, because we are using a PFD, the PLL will lock regardless of our loop filter [2]. Our design made sure that we locked at the desired frequency of 760 MHz. The PLL also generated the harmonics that were expected. This is visible on the spectrum analyzer graph shown below. The *span* feature on the spectrum analyzer can be used to view the actual output frequency and phase noise response over a target range. In this case, the span was 100 MHz.

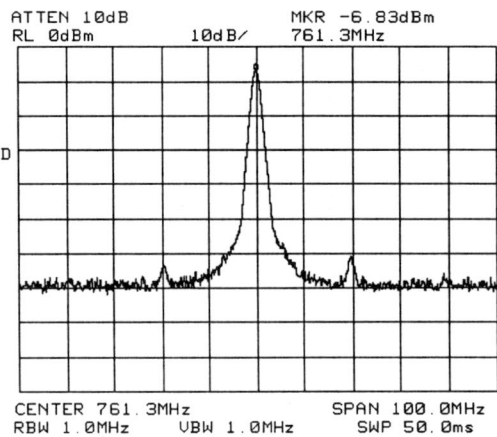

Fig. 4.2.3.2-1. Locked PLL with center frequency of 761.3 MHz and with spurs at multiples of the reference frequency, 19.68 MHz.

Once we got the PLL to lock at the correct frequency, we had to check if our design met its goal. The goal of the design was to design a loop filter, which would get rid of in-phase noise. For this particular IFLO, we needed to provide an output signal of 760 MHz, which we tracked at 761.3 MHz. The Fig. 4.2.3.2-2 illustrates our measurement in phase noise attenuation of approximately –60 dBc/Hz. It was also noted that we used two types of signal generators to simulate our crystal oscillator. We found that the E4433B Agilent ESG-D Series Signal Generator provided a good clean signal similar to a crystal oscillator with very little phase noise added. However, the Hewlett Packard 8648C Signal Generator was measured to provide an additional 4 dB of phase noise to our system. Therefore we decided to work with the Agilent Signal Generator.

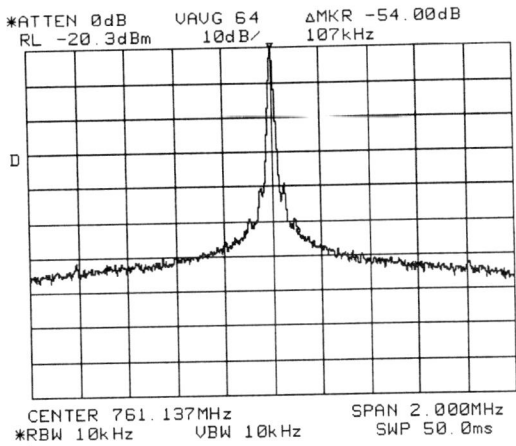

Fig. 4.2.3.2-2. Gives a closer look at the center frequency and the relative close in phase noise. From this figure one can see that we achieved roughly –60 dBc/Hz attenuation.

Next, Fig. 4.2.3.2-3 is a plot of the phase noise measurements at different spot frequencies ranging from 1 kHz to 1 MHz. Note that the amount of attenuation increases linearly with the logarithmic increase in frequency.

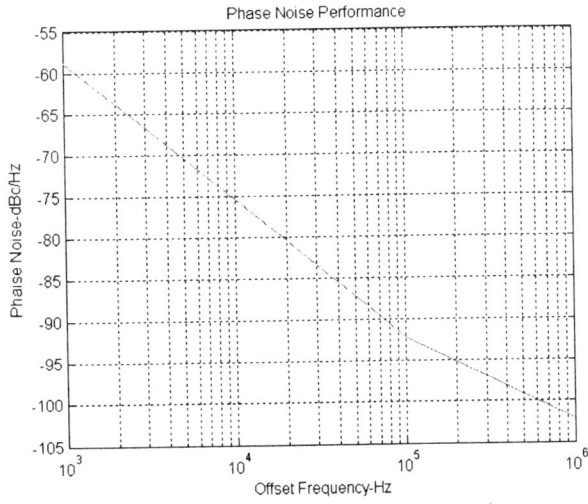

Fig. 4.2.3.2-3. Phase noise performance of our system.

Also, Fig. 4.2.3.2-4 is a graphical representation of the response of our system. This figure indicates that the system has approximately 2 kHz bandwidth. This plot is very similar to the Bode plot that resulted from the simulation, Fig. 4.2.3.2-5.

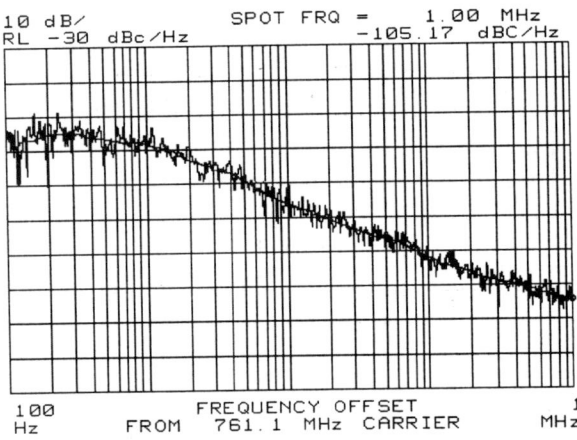

Fig. 4.2.3.2-4. Shows the measured loop bandwidth on the spectrum analyzer.

We simulated our loop performance to verify our design and to compare with the experimental data received. From Fig. 4.2.3.2-5, it can be read that the roll-off frequency in radians is 5.94×10^4. This corresponds to roughly 9 kHz bandwidth. The reason for the differences in bandwidth between simulation and experimental results are explained in Section 4.2.3.3.

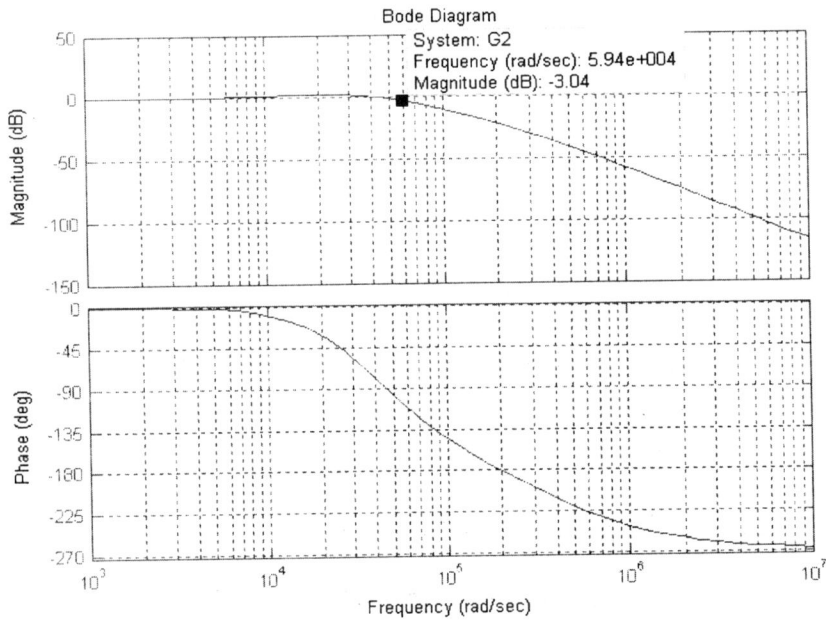

Fig. 4.2.3.2-5. Simulation of loop bandwidth performance.

4.3. Design Comparison / Design Trade Off

4.3.1. Receiver

We have already specified the required performance of our receiver and the filter. The amplitude response over frequency is one of the most important aspects of our design. It is crucial to the overall system performance in that it determines the selectivity of our filter. It is a necessary factor in reproducing the original transmitted signal by avoiding signal distortion. This sets the allowable tolerance or dynamic range of our system. The frequency response is very close to those of the preliminary design in *PSpice*. There are notable similarities between Fig. 4.3.1-1 and Fig. 4.3.1-2. As expected the pass-band is maximally flat, giving a cutoff and corner frequency at 3 MHz.

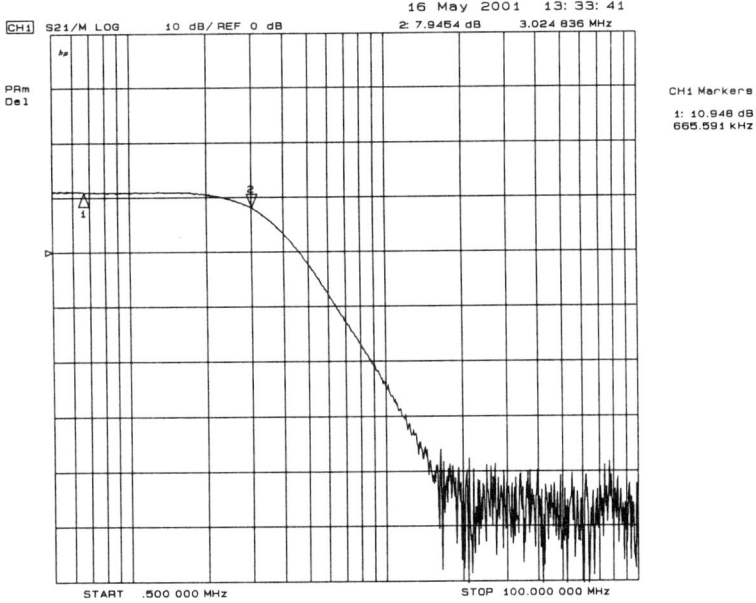

Fig. 4.3.1-1. Actual frequency response of the low-pass filter.

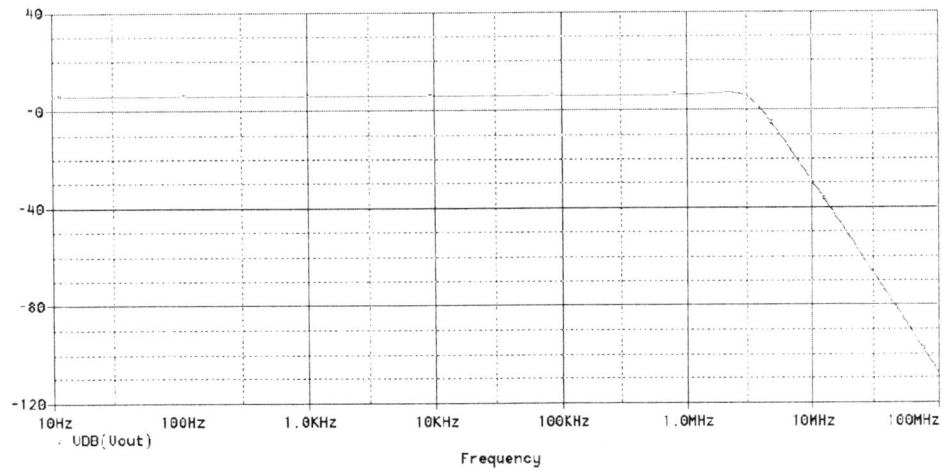

Fig. 4.3.1-2. Frequency response of the low-pass filter (*PSpice* simulation).

Poles and positions were also looked at and were designed using the Sallen-Key method to give the maximally flat response. We can see from Fig. 4.3.1-3 that about half-way to the 3 dB point, group delay departs from flatness and rises to a peak near the 3 dB point. The larger the number of poles, the more rapid the amplitude drop-off beyond the 3 dB point and the more the group delay deviates from flatness. The data presented in Fig. 4.3.1-3 and Fig. 4.3.1-4 represent the group delay response and phase response of our Butterworth low-pass filter.

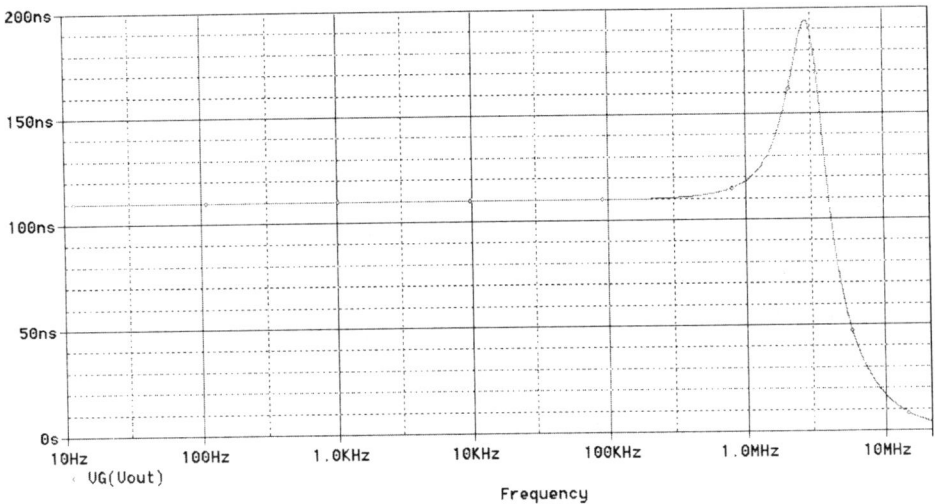

Fig. 4.3.1-3. Group delay of the Butterworth low-pass filter.

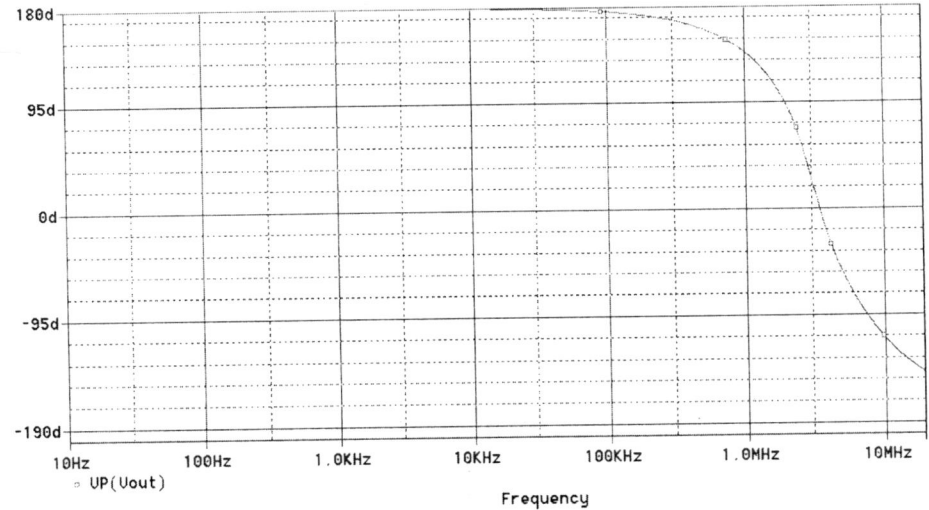

Fig. 4.3.1-4. Phase response the Butterworth low-pass filter.

The actual group delay and phase response of our filter generated by the network analyzer is similar to the one simulated in *PSpice*. The circuit was "tweaked" to show smaller deviation by replacing some of the components in order to obtain desired results. A complete list of components and values showing design changes to improve performance is given in Table 4.2.1-2.

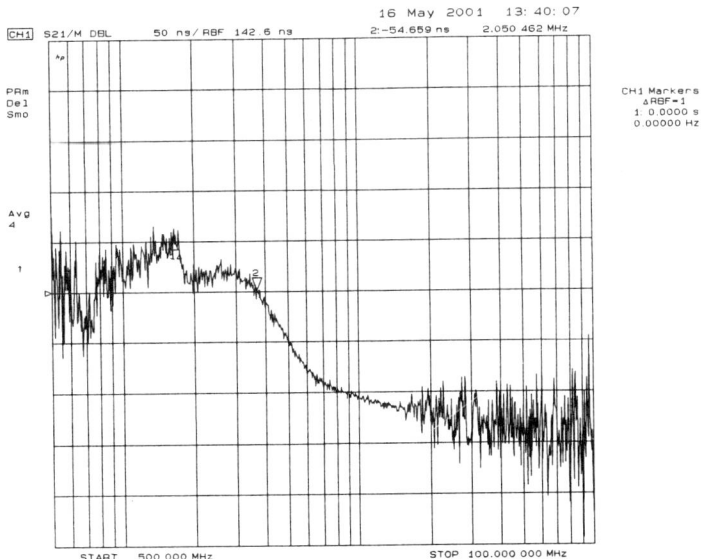

Fig. 4.3.1-5. Group delay of the Butterworth low-pass filter.

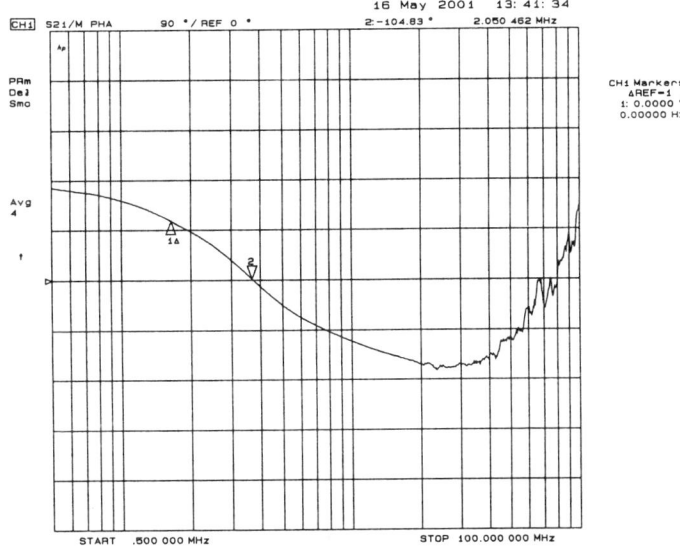

Fig. 4.3.1-6. Phase response of the Butterworth low-pass filter.

The various characteristics we have presented (amplitude and phase) are important since relative delays between components can introduce intersymbol interference. So it is desirable to have a rise time that allows the signaling element to attain its ultimate amplitude before the next element is received. Adjusting the locations of the poles and zeros of the transfer function is one way of controlling the rise time so that maximum amplitude is obtained. For this reason and reasons explained earlier, it is necessary to replace and change the components from the actual theoretically calculated values. For example, although the group delay in the *PSpice* model clearly shows the normal spike, the delay is not so easily seen as in the actual response generated by the network analyzer.

4.3.2. Transmitter

Due to the effects mentioned above, and the combined effects of component tolerances, we cannot expect our initial response to identically match the theoretical response. Indeed, as shown in Fig. 4.3.2-1 below, the initial magnitude response was centered 10 MHz lower than desired.

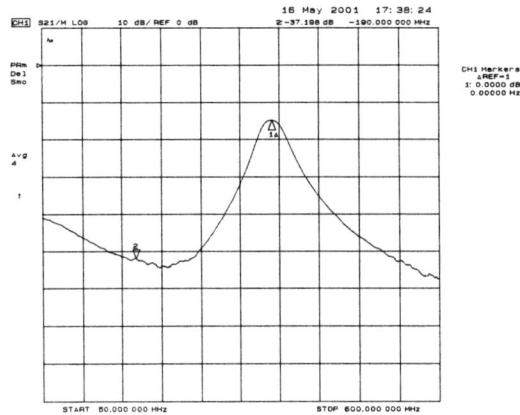

Fig. 4.3.2-1. Initial magnitude response of prototype.

To compensate, the capacitances C1 and C2 were decreased in order to raise the center frequency. After several tries using values between 8 and 12 pF, a response centered around the desired frequency of 380 MHz was obtained, as shown in Fig. 4.3.2-2.

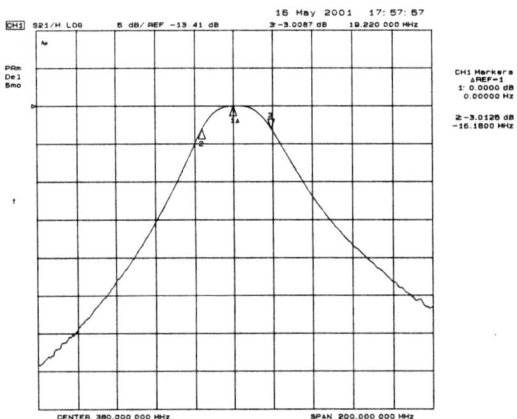

Fig. 4.3.2-2. Filter response after tuning.

From this plot the prototype filter has a 3 dB bandwidth of 35.4 MHz (363.8 to 399.2 MHz), and the insertion loss appears to be 13.41 dB. However, we must subtract the loss introduced by the baluns and resistors used to match the circuit to the 50-Ω single-ended network analyzer, yielding an insertion loss comparable to if not better than that of the standard SAW / IF filter.

We are also interested in knowing the attenuation of signals at our critical frequencies of 190 MHz and 760 MHz. As is clear from Fig. 4.3.2-3 and Fig. 4.3.2-4, these attenuations are

39.7 and 44.2 dB, respectively. Both values exceed the typical attenuation of the standard SAW filter of 37 and 40 dB (respectively) at these frequencies.

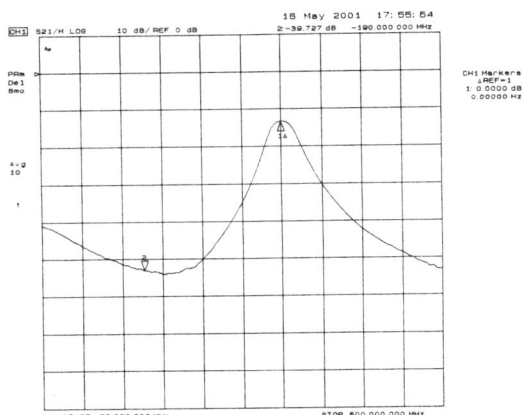

Fig. 4.3.2-3. Measured filter response, 50 – 600 MHz.

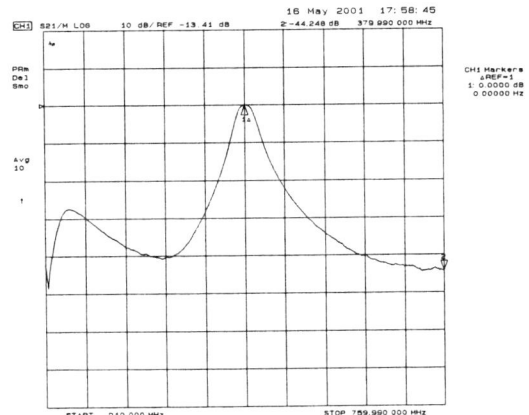

Fig. 4.3.2-4. Measured filter response to 760 MHz.

Finally, we compare the final measured response with the response obtained by simulation, using real-valued components. The response shown in Fig. 4.3.2-5 has a 3 dB bandwidth of 41 MHz (355 to 396 MHz), with an insertion loss of nearly 12 dB. It may seem surprising at first that the measured response is slightly *better* than the simulated response; however, this is easily explained. The values included in the simulation circuit to compensate for parasitic capacitance and finite inductor Q must have been slightly pessimistic. In addition, the measured response does not decrease steadily as in the simulation but shows a large, gentle ripple at lower frequency. Fortunately this anomaly occurs in a non-critical location (at which no significant spurious mixing products are expected) and for our purposes it can safely be ignored.

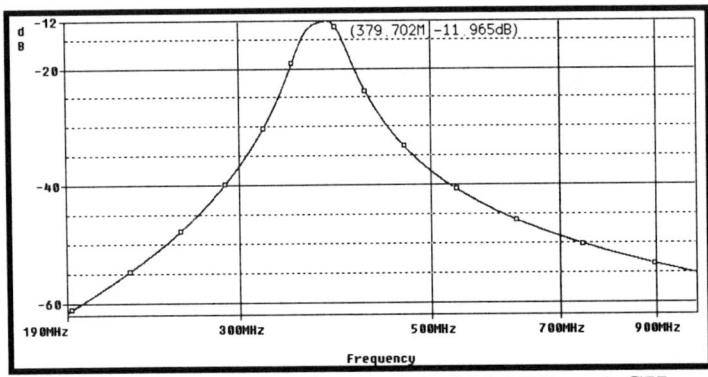

Fig. 4.3.2-5. Simulated response, 190 MHz – 1 GHz.

4.3.3. Frequency Synthesizer

The results of both simulation and the experiment have been stated and there are differences. The scope of this section is to explain the origin of these discrepancies. Perhaps the most noticeable difference is the system bandwidth, which is a function of the input and output parameters. Dissimilarities in the input parameters, namely VCO gain and current gain, most definitely cause the output parameters, or resistor and capacitor values and bandwidth, to differ from what was to be expected. For clarity all values, input and output, are presented in tabular form.

Table 4.3.3-1. Simulation and experimental data.

Parameter	Simulation Value	Experimental Value
K_{vco}	15 MHz/V	30 MHz/V
I_{cp}	2.1 mA	175 µA
F_{ref}	75 kHz	75 kHz
F_{out}	760 MHz	761.13 MHz
N-divider	10133.33	10133
C_1	227.69 pF	220 pF
C_2	81660 pF	68000 pF
C_3	22.769 pF	Open
R_2	11.332 kΩ	11 kΩ
R_3	406.43 kΩ	Short
Settling Time	230 µs	5 ms
Bandwidth	9.45 kHz	1.5 kHz

Actual VCO gain and I_{cp} gain were measured in the lab to see if they differed from what was expected. There was a sizable difference between what was measured and what the design had anticipated. Table 4.3.3-1 indicates that the current gain in the experiment is much less than what was expected. The effect of this decrease in current is a much smaller bandwidth, which can cause degradation in phase noise performance and slow the settling time of the system. The increase in VCO sensitivity in the experiment can also slow the settling time of the system and allow thermal noise from the loop filter to find its way onto the VCO output [15]. With the knowledge of these key parameters being different from what were designed for, the entire set of calculations used to find RC values is compromised.

Another justification for our experimental results stems from the fact that implementing our exact capacitor and resistor values is not possible. For production reasons, capacitors and resistors are quantized according to 5% values that are standard in industry. We also had to quantize our values according to the industry standard, Table 4.2.3.1-1. It should be noted that the experimental comparison frequency is slightly different from the simulated value and this is because the computer program that controls the N-divider, in the experiment, can only take in whole numbers. However the result of this difference is negligible compared to the effects of K_{vco} and I_{cp}.

Lastly, the EV board that was used was not set up for a third-order system. That is, the pad for C_3 was left open and the pad for R_3 was shorted. In order for us to implement our third-order system we would have had to cut traces and risk permanently damaging the board. This resulted in our experimental results being based on a second-order system, where our design was for a third-order system.

Based on the approval of Mr. Devries, our Maxim advisor, our design was on the right track and the values for the first trial were better than expected. Had we had an additional workbench (all teams shared the same bench) we could have been able to make adjustments to some parameters and components, which were the sources of noise that we had identified.

Chapter Five – Administrative

5.1. Introduction

The administrative portions of the design project can be divided into two categories, budgeting and scheduling, respectively. These administrative portions are essential in planning the execution of each task throughout the project lifetime properly. Budgeting is obviously a primary concern as it enables us to plan the financial aspects of the project to ensure that the financial means are available for completion of the project. Scheduling is absolutely necessary to coordinate meetings and communicate progress amongst pertinent personnel.

5.2. Cost Analysis

The procedure to calculate cost and to write a business plan in this project is exactly the same as that outlined in class. However, this design is a part of a previously produced item at Maxim. Thus, most relevant expenses such as marketing and advertising are already paid through the earlier version of this item. The production cost will not change from the earlier version. Therefore we do not include that here either. The following shows the profit that this design will bring to the enhanced version of this product.

Prototype Cost
    Parts $530
    Software $2,920
    Direct Labor (6 personnel x 12 hours per day x 2 days x $25 per hour) $3,600
    Capital Equipment (purchase price = $94,045; but equipment on-hand) $0
    (depreciation schedule: 40% first year, 30% second & third year)

Estimated Costs to Produce 500,000 Units
    Parts (Purchased at Bulk Rate) $796,360
    Labor (Assuming 10 units per hour @ $25 per hour) $0
    Storage ($0.70 per square foot per month for 300 Sq. Ft. for 6 Months) $0
Advertising Costs $0
Cost (sub-total) $803,410
Overhead Cost (5% of Total) $40,170
TOTAL COST $843,580

Expected Sales   490,000 Units
Estimated Price Per Unit (All filters combined) $1.72
Current Price Per Unit (All filters combined) $2.00
Savings Per Unit $0.28
TOTAL SAVINGS $137,200

Hiring Bonus Cost ($10,000 per person) $60,000

EXPECTED TOTAL PROFIT FOR MAXIM $77,200

5.3. Organization of the Task

The project tasks are divided and distributed to each of the three teams: receiver team, transmitter team, and frequency synthesis team. The division of tasks between teams allows us to specialize in different areas of the W-CDMA transceiver to provide maximum efficiency towards the tasks at hand. This method of task division is an efficient means to accomplish a large task in a small amount of time; however, the true meaning of teamwork becomes apparent when the project end nears and each team member uses their individual brush to paint their portion of the final mosaic.

The primary means of scheduling tasks throughout the project are e-mail, teleconferences, meetings, and lectures. Logs of all communications are kept in each of our respective notebooks for record keeping. During the initial phase of the project, scheduling consists of the coordination of research assignments with our respective mentors via e-mail in order to guide us in the development of the necessary radio frequency design skills. Scheduled lectures given by mentoring Maxim engineers helped fine-tune the skills we are developing. Upon development of these skills, tasks were assigned from our mentors via e-mail to calculate necessary radio frequency design parameters pertinent to our respective designs. Teleconference sessions were established to answer technical questions that arose throughout the project. The communication links utilized to schedule the hardware design, implementation, and test phase of our project in Sunnyvale were strictly routed via the project advisor. A summary of the project timelines is provided in Table 5.3-1.

Table 5.3-1. Project timeline.

January	Senior design projects assigned
	Problem definition
	Purchase RF Microelectronics textbook by Ben Razavi
	Begin to read text and research on internet
February	Read first five chapters of Razavi
	Bob Kelly lectures on basic RF electronics
	Assignments to teams with respective mentors
March	Analysis of reference design begins
	Research continues; RF system analysis begins
	Begin preparing for mid-quarter presentations
April	Dave Devries lectures on effects of noise in a transceiver
	Mid-quarter presentations
	Focus shifts from "RF analysis" to "filter design"
May	Extensive research in filter design
	Filter design simulations
	Preparation for trip to Maxim Laboratories in Sunnyvale, CA
	Design, implementation, & testing of prototype filters
	Post-implementation analysis
June	Complete final report
	Generate slides for oral presentation

Chapter Six – Meeting Expectations

6.1. Introduction

This chapter focuses on the issue of whether or not the initial specifications of the design have been met. Perhaps the most substantial part of any project lies in the results that show if the design works. Section 6.2 explicitly declares whether the design has satisfied all the desired specifications. Section 6.3 summarizes important elements of design.

6.2. Design Constraints

As the constraints are very different for each of the three subproblems, they are addressed separately in the following three subsections.

6.2.1. Receiver

Many designs are created in today's world, however there must sometimes be some sacrifices. These include values on actual components not meeting designed component values, other component selections that may be suited to do the job but have features that are unnecessary, and other related hardware that is used for one design but not suitable for the task at hand.

When manufacturers create components such as resistors and capacitors, it is impossible to produce every value from one to infinity. For example, resistors come in discrete values like 1 kΩ, 2 kΩ, and 10 kΩ. Such is the case when building the physical prototype of our design. In Table 6.2.1-1 the designed values for resistors and capacitors are compared with the actual values used. One of the actual capacitors varied by 6 pF from the designed value. While the designer can take into account the fact that these components are not available for every value he or she can imagine, a certain amount of common sense as well as experience play a major role in the selection of components. Our design required some "tweaking" in order to correct for a bump located at the corner frequency. In the *PSpice* simulations a bump created an unwanted extra gain in the pass-band before transitioning to the stop-band as illustrated in Fig. 4.3.1-2. For this, the resistor and capacitor values of the second stage were altered to provide a smoother transition and to flatten out the bump, Fig. 4.3.1-1.

Table 6.2.1-1. Comparison between designed component values and actual values used (continued on the next page).

Component	Designed Value	Actual Value
C_1	74 pF	68 pF ($\pm 5\%$)
C_2	21 pF	22 pF ($\pm 5\%$)
C_3	22 pF	22 pF ($\pm 5\%$)
C_4	72 pF	68 pF ($\pm 5\%$)
R_1	1 kΩ	1.07 kΩ ($\pm 1\%$)
R_2	329 Ω	332 Ω ($\pm 1\%$)
R_3	2.3 kΩ	2 kΩ ($\pm 1\%$)

R_4	1 kΩ	1.07 kΩ (±1%)
R_5	1 kΩ	1.07 kΩ (±1%)
U_1	THS4121	THS4121
U_2	THS4121	THS4121

In our design we call for a single-ended output from our filter, yet we use an op-amp, which supports differential outputs. The op-amp is within the range of performance that is required by our design and meets these specifications, but is over-designed because it supports a feature that is not used. This design constraint arose from the fact that the op-amp used was the only one available to us that also met our design specifications.

The last design constraint that affected our design was the etched surface-mount board that was used to build our designed circuit. In the interest of time, the board used was designed for another filter, but had some similar design elements. For example, the circuit that the board was etched for used two op-amps. Since our designed circuit used two op-amps with surrounding resistors and capacitors, the board was suitable. It is important to note that none of these constraints had any adverse affect on our filter's performance.

6.2.2. Transmitter

The transmitter BPF design specifications in Table 2.3.2-1 require the IF / BPF to exhibit a center frequency of 380 MHz, a 3 dB bandwidth of at least 6.8 MHz, at least 37 dB of attenuation at 190 MHz, and at least 40 dB of attenuation at 760 MHz. Without taking finite inductor Q into account, designing a filter on paper to meet specifications is relatively easy. However, at our frequencies, once the effects of finite Q are taken into account, the insertion loss and 3 dB bandwidth can change drastically.

The availability of discrete-valued components (mentioned in the previous section) and the presence of stray capacitance are also constraints. However, these constraints can be partially overcome using the method to fine-tune a resonant circuit discussed in Section 4.3.2. This constraint of finite component values can be insightful to a beginning filter circuit designer, as the process of shaping the response of the filter immediately demonstrates practical effects of various component changes.

Perhaps a more crucial constraint is that of time during the implementation and testing phase. If the time available to assemble and test a high-frequency circuit is limited, as it usually is, this constraint proves very difficult to overcome. Many time-consuming circuit modifications can be necessary to assemble, test, modify, and fine-tune a filter design. This constraint is evident in Section 4.2.2, where many modifications become necessary prior to obtaining a satisfactory filter response.

Last, but certainly not least, financial constraints must be considered. Based on the cost analysis in Section 5.2, a production of 500,000 filters at a successful sales rate of 98% will result in an expected total profit of $77,200. This amount of profit results after a hiring bonus of $10,000 is given to each of the six graduating engineers assigned to the project. With all circuit constraints, time constraints, and financial constraints considered, the experimental test results in Section 4.3.2 confirm that the filter design specifications have been met.

6.2.3. Frequency Synthesizer

The major goals in the design of a PLL frequency synthesizer, aside from producing the desired signal at a determined frequency, are to achieve low phase noise, low spurious output, and a quick settling time. The results for all three of these aspects have already been determined and presented in this report for both simulation and experiment. This section discusses whether the results meet industry requirements, and if not then what changes need to be made.

Phase noise:

Table 6.2.3-1. A typical VCO phase noise response [20].

Offset (kHz)	Phase noise (dBc)
1	-80
12.5	-105
30	-111
120	-121
900	-128

Our phase noise results:

Table 6.2.3-2. Measured VCO phase noise response.

Offset (kHz)	Phase noise (dBc)
1	-58.3
10	-75.3
100	-92.17
1000	-102.17

By comparing the above tables, it appears that our results are a little noisy. This is largely due to the small bandwidth having little effect on VCO phase noise. An alternative approach to the design, Fractional-N architecture, may have achieved better phase noise performance, but the degradation in spurious noise performance may not have been a wise tradeoff. Our spurious noise attenuation level at the first offset reference multiple is about –58 dB. For the transmit IF / PLL, the spurious noise level requirement is –58.16 dB [5]. Our spurious noise attenuation is acceptable but the overall phase noise performance did not fully meet minimum noise requirements. Any extra noise that is present from 0 to 1.5 kHz offset can be contributed to errors in loop filter design and has been justified in Section 4.3.3. The very same elements that cause our experiment to stray from our simulation cause our experiment to have a less than desired phase noise performance. Any extra noise present outside of 1.5 kHz is dominated by tank components, like the inductor and varactor, which are related to but not part of our design. Better attenuation can be achieved by using tank components with a higher quality value [11]. Other possible reasons for insufficient noise performance outside our design reside in the bad board layout resulting in a crosstalk, and noise from the voltage supply.

The experimental settling time of our design was 5 ms. Under normal operating conditions for an IF / PLL, this time is sufficient since it does not need to switch from channel to channel. However, for a time duplex mode, the PLL would be turned off to conserve power, then turned on again to transmit the signal. This would create the need for a fast settling time of no more than 5 ms.

This design worked but could benefit tremendously from a few modifications. With the knowledge of actual design parameters, a revised design can be tailor-made to fit the system. The new design would have a profound increase in phase noise performance due to the customization of the loop filter to the surrounding components, whose exact characteristics are now known.

6.3. Elements of Design

Issue	Effects of our design
Economic Factors	There are many economic factors that influence a design. We have covered a cost analysis in Chapter Five, which gives us a good idea of what it would cost to mass-produce our design. Though this is a good rough estimate, the fact that there are other companies out there wanting to make money from producing a similar product is another factor that must be considered. Through supply and demand, many companies may not be competitive enough to survive and this would cause loss of jobs, however it would also supply the customer with superior products.
Safety	While the actual manufacturing processes involved with mass-producing our design may involve some drawbacks to the environment, all responsibilities fall on the manufacturer to conduct itself accordingly. As for our responsibilities when making a prototype, all known precautions were taken to ensure the safety of all involved.
Reliability	No extensive tests were made to ensure the long-term reliability of our design. However, to the best of our knowledge, we have no reason to believe that our design would not be worthy of long-term use.
Aesthetics	Prototypes that were made during this project were constructed for test purposes only and were not optimized for appearance. They were merely meant to provide us with knowledge and skills related to our particular design. With further design and testing a prototype worthy of being produced, which not only works to designed specifications, but also looks good, could be achieved.
Ethics	In reviewing the IEEE Code of Ethics, we have done nothing but try to better our understanding of this technology as well as help each other along the way, working together as a team. We often criticized and complimented each other during our time working together. Throughout this project, we have conducted ourselves in nothing but a professional manner and have done nothing to harm the reputations of others involved in this project. By reporting all of our work and findings in this report, we have given credit where needed and we have presented our findings with the utmost honesty.
Social Impacts	With many people desiring portable communications devices in today's fast-paced world, our design will play a major role in furthering the wants and needs of these types of people.

Chapter Seven – Conclusions

7.1. Introduction

This chapter is designed to give closure to the entire project and report. Each of our three design tasks proved to be challenging and to require creativity. Intriguing hours were spent on understanding each component and its effect on the entire system. Our designs proved to be accurate through simulation and by way of actual implementation, and can be implemented in industry with only minor modifications. We are satisfied with our results and feel that we have fulfilled our job requirements.

7.2. Expansion and Improvement

7.2.1. Receiver

Due to a time constraint certain aspects like temperature were not tackled in our design. These aspects of design were not crucial to a successful demonstration of our design. Another consideration that was not addressed was the concept of "commercial application." Therefore, this report creates the foundation for future investigations in those areas. Although some of these issues were not considered, those did not impact our design in any way. The issues that were not addressed could be something for future expansion and design improvement for the next generation. Overall, the results presented in this report regarding all aspects of our design show that our design worked quite well.

7.2.2. Transmitter

The UCR / Maxim Senior Design Project has been a very valuable learning experience. The project has given us valuable experience in radio frequency communication system analysis and design, which would not have been obtained through available coursework at UCR. The project has provided us with an exposure to "real world" engineering problem solving techniques and has helped us to establish networking foundations in the corporate environment. Some possibilities for expansion within the transmitter project are to examine in greater detail one of the other areas within the transmit path, such as the quadrature modulator, variable gain amplifiers, or power amplifier. A possibility for future work on the IF / BPF is to implement the filter on-board a working cellular phone in order to determine its practical advantages and disadvantages.

7.2.3. Frequency Synthesizer

Though this project has resulted in a complete working design, there is much to be desired and that can be passed on to the next group of students that wish to take on this project. Since this was a new project, new goals and objectives were created as the project developed. The next group of students could benefit from a pre-developed time line that gives explicit tasks and dates for which each task should be completed. This would ensure proper progress throughout the project. Indefinite access to the proper test equipment would tremendously improve the end results because the students would have the capabilities to design and test as they go through each step in building their project.

Future students should also look into implementing the third-order loop filter, which would provide a spurious-free signal. The design of the third-order filter is complete, so future students may want to take measurements and compare this to the simulated and actual requirement in order to analyze its impact on the system. Future students should pay closer attention to the amount of charge current being used so that techniques such as turbo tracking can be implemented. Understanding and knowing how the equipment works will speed up the process.

7.3. User's Manual

The MAX2310 EV kit is used in testing the receive low-pass filter, and the MAX2360 EV kit is used in tests involving the transmit IF band-pass filter and frequency synthesizer designs. Please refer to the corresponding evaluation kit documentation for additional information on conducting tests with these boards.

Appendix A: Parts List

A.1. Receiver portion

Part Name	Quantity	Value/Part Number	Manufacturer
Resistor	3	1.07 kΩ (±1%)	
Resistor	1	2 kΩ (±1%)	
Resistor	1	332 Ω (±1%)	
Capacitor	2	68 pF (±5%)	Murata
Capacitor	2	22 pF (±5%)	Murata
Capacitor	4	0.1 μF (±5%)	Murata
Op-Amp	2	THS4121	Texas Instruments
Jumper Wires	6	Various lengths	
Surface Mount Board	1		
Coaxial Connectors	3		

A.2. Transmitter portion

A.2.1. Production circuit (given in Fig. 3.2.2-3)

Designation	Qty	Value	Part Number
C_1	1	12 pF 5% ceramic capacitor (0402)	Murata GRM36C0G120J050
C_2	1	10 pF 5% ceramic capacitor (0402)	Murata GRM36C0G100J050
C_{12a}, C_{12b}	2	1.6 ± 0.1 pF ceramic capacitors (0402)	Murata GRM36C0G1R6B050
L_{1a}, L_{1b}	2	15 nH 5% inductor (0603)	Coilcraft 0603CS-15NXJBC
L_2	1	6.8 nH 5% inductors (0603)	Coilcraft 0603CS-6N8XJBC

A.2.2. Prototype circuit (given in Fig. 4.2.2.3-1)

Designation	Qty	Value	Part Number
C_{1a}, C_{1b}	2	8 ± 0.5 pF ceramic capacitor (0402)	Murata GRM36C0G080D50
C_{2a}, C_{2b}	2	1.5 ± 0.1 pF ceramic capacitors (0402)	Murata GRM36C0G1R6B050
C_3, C_4	2	1.5 ± 0.1 pF ceramic capacitors (0603)	Murata
L_1, L_2	2	15 nH 5% inductors (0805)	Coilcraft
R_1, R_2	2	200 Ω 5% resistors (0603)	
R_3, R_4	2	100 Ω 5% resistors (0603)	
T_1, T_2	2	Balun transformers (B5F type)	Toko 458DB-1011
	2	SMA connectors	Digikey J502-ND
	1	PCB	
	6"	Coaxial wire	

A.3. Frequency synthesizer portion

Designation	Qty	Value	Manufacturer
C_1	1	220 nF 5% ceramic capacitor	Murata
C_2	1	68 pF 5% ceramic capacitor	Murata
R_2	1	11 kΩ 5% resistor	

Appendix B: Equipment List

HP 8560E Spectrum Analyzer (30 Hz – 2.9 GHz)	$19800
Agilent E4433B ESG-D Series Signal Generator (250 kHz – 4 GHz)	$18450
Hewlett Packard 8753E RF Network Analyzer (30 kHz – 6 GHz)	$35600
Hewlett Packard E3630A Triple Outlet DC Power Supply	$525
Hewlett Packard 8648C Signal Generator (100 kHz – 3.2 GHz)	$16000
Personal Computer	$2000
Stereo Microscope (20X)	$1450
Weller WES50 Soldering Station (2 @ $110 ea)	$220

Appendix C: Software List

MATLAB, Version 6.0	$845
MAX 236x EV PC-Control Software Ver 5.00.26	Freeware
Microsoft Office XP Professional	$580
ORCAD PSPICE, Version 9.0	$1495
PLL Made Easier, Version 1.5	Freeware

Appendix D: Special Resources

Maxim Headquarters Facilities
Maxim Technical Personnel

Appendix E: Parts Cost Breakdown

MAX 2361 Evaluation Board $200
MAX 2310 Evaluation Board $200
Printed Circuit Board $30

Transmitter Differential Band-Pass Filter (goes on EV board or a production unit):
(2) 6.8 nH 5% inductors (0603)   10ea = $5.63, 1000ea = $198.75
(1) 15 nH 5% inductor (0603)   10ea = $5.63, 1000ea = $198.75
(2) 1.6 pF ± 0.1 pF ceramic capacitors (0402)   10ea = $2.38, 1000ea = $91.50
(1) 10 pF 5% ceramic capacitor (0402)   10ea = $0.77, 1000ea = $18.20
(1) 12 pF 5% ceramic capacitor (0402)   10ea = $0.77, 1000ea = $18.20

Transmitter Band-Pass Filter Prototype:
(2) 1.5 pF ± 0.1 pF ceramic capacitors (0603)   10ea = $0.87, 1000ea = $32.30
(2) 9.5 pF 5% ceramic capacitors (0402)   10ea = $0.77, 1000ea = $18.20
(2) 15 nH 5% inductors (0804)   10ea = $5.63, 1000ea = $198.75
(2) 100 Ω 5% resistors (0603)   10ea = $0.80, 1000ea = $16.82
(2) 200 Ω 5% resistors (0603)   10ea = $0.80, 1000ea = $16.82
(2) Balun transformers (B5F type, TOKO 458DB-1011)   1ea = $2.89, 100ea = $186.12
(2) SMA connectors (DIGI-KEY J502-ND)   1ea = $5.88, 500ea = $1265.25
(1) Coaxial wire spool   100ft = $34.39, 500ft = $150.46

Receiver Low-Pass Filter:
(1) 332 Ω 1% resistor (0402)   10ea = $0.84, 1000ea = $177.00
(3) 1.07 kΩ 1% resistors (0402)   10ea = $0.84, 1000ea = $177.00
(1) 2 kΩ 1% resistor (0402)   10ea = $0.84, 1000ea = $177.00
(4) 0.1 uF ± 5% ceramic capacitors (0603)   1ea = $0.24, 10000ea = $1000.00
(2) 22 pF ± 5% ceramic capacitors (0603)   1ea = $0.24, 10000ea = $1000.00
(2) 68 pF ± 5% ceramic capacitors (0603)   1ea = $0.24, 10000ea = $1000.00
(2) THS4121 Texas Instruments Op Amp (DGN)   1ea = $3.51, 5000ea = $9750.
(1) Jumper Wire 30 AWG   75ea = $5.49, N/A
(1) Surface Mount Board   1ea = $3.08, N/A
(3) SMA connectors (DIGI-KEY J502-ND)   1ea = $5.88, 500ea = $1265.25

Frequency Synthesizer Loop Filter:
(1) 11 kΩ 5% resistor (0402)   10ea = $0.80, 1000ea = $16.82
(1) 68 nF ± 5% ceramic capacitors (0402)   1ea = $0.24, 10000ea = $1000.00
(1) 220 pF ± 5% ceramic capacitors (0402)   1ea = $0.24, 10000ea = $1000.00

Appendix F: *Matlab* Code for Loop Filter Design

```matlab
%Heath Deuel
%Oscar L. Servin
%Senior Design

%Software to design a loop filter.

Phase = 55;                 %Degrees.
Kvco  = 15e6;               %Hz/V.
Kphi  = 175e-3;             %Amps.
Fref  = 75e3;                       %Hz.
Fout  = 760e6;                      %Hz.
ATTEN = 20;                         %dB.

N = Fout/Fref;

%First time constants.
T1 = (sec(Phase*pi/180)-tan(Phase*pi/180))/1.256E5;

%Third time constants.
T3 = sqrt((10^(ATTEN/20)-1)/(2*pi*Fref)^2);

%Cutoff frequency.
wc1 = (tan(Phase*pi/180)*(T1+T3))/((T1+T3)^2+(T1*T3));
wc2 = (((T1+T3)^2+(T1*T3))/((tan(Phase*pi/180)*(T1+T3))^2));
wc3 = sqrt(1+wc2)-1;
wc = wc1*wc3        %rad
BW3 = wc/(2*pi)     %Hz

%Second time constant.
T2 = 1/((wc^2)*(T1+T3));

%Second-order bandwidth.
wp = 1/(sqrt(T2*T1))    %rad
BW2 = wp/(2*pi)         %Hz


C1 = (T1/T2)*((Kphi*Kvco)/(wc^2*N))*(sqrt((1+wc^2*T2^2)/((1+wc^2*T1^2)*(1+wc^2*T3^2))))
C2 = C1*(T2/T1 - 1)
C3 = C1/10

R2 = T2/C2

%Rule of thumb, choose R3 to be AT LEAST twice the value of R2!
R3 = 0.1*(C2*R2)/C3

%Damping ratio.
zeta = ((R2*C2)/2)*sqrt((Kphi*Kvco)/(N*(C1+C2+C3)));

%Natural frequency.
wn = sqrt((Kphi*Kvco)/(N*(C1+C2+C3)));

%Settling time.
Ts = 4/(zeta*wn)
```

```
pause

C1 = 330e-12;
C2 = 68000e-12;
R2 = 11e3;

%Transfer function second-order approximation (closed loop).
numZ = [(C2*R2) 1];
denZ = [(C1*C2*R2) (C1+C2) 0]

Z = tf(numZ,denZ)
H = tf([1],[N 0])
G = H*Z*Kvco*Kphi

%Third-order open loop.
numZ1 = [(C2*R2) 1];
denZ1 = [(C1*C2*C3*R2*R3) (C2*C3*R2+C1*C2*R2+C1*C3*R3+C2*C3*R3) (C1+C2+C3) 0]

Z1 = tf(numZ1,denZ1)
H1 = tf([1],[N 0])
G1 = H*Z1*Kvco*Kphi

%Third-order closed loop.
numZ2 = [(C2*R2) 1];
denZ2 = [(C1*C2*C3*R2*R3) (C2*C3*R2+C1*C2*R2+C1*C3*R3+C2*C3*R3) (C1+C2+C3) 0]

Z2 = tf(numZ2,denZ2)
H2 = tf([1],[N 0])
G2 = H*Z2*Kvco*Kphi/(1+H*Z2*Kvco*Kphi)

clf
grid on
Bode(G)
hold on
%Bode(G1)
%Bode(G2)
```

References

[1] D. Banerjee, "PLL Performance, Simulation, and Design," [Online document], Available in http://www.national.com/appinfo/wireless/deansbook.pdf (current Sept. 12, 1993).

[2] C. L. Barker, "Introduction to Single Chip Microwave PLLs," [Online document], Available in http://www.national.com/an/AN/AN-885.pdf.

[3] R. Best, *Phase-Locked Loops: Design, Simulation, and Application*. New York: McGraw-Hill, 1999.

[4] C. Bowick, *RF Circuit Design*. Boston, MA: Newnes, 1997.

[5] A. Franceschino, "Phase Locked Loop Primer and Application to Digital European Cordless Phone," *Applied Microwave & Wireless*, pp. 65-84, Fall 1994.

[6] K. Holladay, "Design Loop Filters for PLL Frequency Synthesizers," [Online document], Available in http://www.mwrf.com/1999/sep1599/098-104.html (current Sept. 1, 1999).

[7] L. P. Huelsman, *Active and Passive Analog Filter Design: An Introduction*. New York: McGraw-Hill, 1993.

[8] W. O. Keese, "An Analysis and Performance Evaluation of a Passive Filter Design Technique for Charge Pump Phase-Locked Loops," [Online document], Available in http://www.national.com/an/AN/AN-885.pdf (current May 26, 1996).

[9] B. Kelly, "UC Riverside / Maxim W-CDMA Project," PowerPoint Slides, [Online document], Available in http://www.maxim-ic.com (current Feb. 14, 2001).

[10] K. Lacanette, "A Basic Introduction to Filters-Active, Passive, and Switched-Capacitor," [Online document], Available in http://www.national.com/an/AN/AN-779.pdf (current May 30, 2001).

[11] H. C. Luong, W. S. T. Yan, "A 2-V 900-MHz Monolithic CMOS Dual-Loop Frequency Synthesizer for GSM Receiver," *IEEE Journal of Solid-State Circuits*, vol. 36, no. 2, pp. 204-216, Feb. 2001.

[12] O. K. Jensen, T. E. Kolding, C. R. Iversen, S. Laursen, R. V. Reynisson, J. H. Mikkelsen, E. Pedersen, M. B. Jenner, "RF Receiver Requirements For 3G W-CDMA Mobile Equipment," *Microwave Journal*, pp. 22-46, Feb. 2000.

[13] M. L. Jewell, "On High Frequency Narrow-Band Elliptic Filters," *IEEE Trans. Circuits Systems*, vol. 37, no. 2, pp. 264-267, Feb. 1990.

[14] B. Razavi, *RF Microelectronics*. Upper Saddle River, NJ: Prentice Hall, 1998.

[15] M. Smith, *An improved PLL design method ω_n and ξ*, Tech. Report TN-98011500, Motorola Semiconductor, Denver, CO, 1998.

[16] K. L. Su, *Analog Filters*. London: Chapman & Hall, 1996.

[17] B. Trump, M. Stitt, "Sallen-Key Low-pass Filter Design Program," Available in http://www.burr-brown.com/download/Abs/AB-017.pdf (current Aug. 1991).

[18] A. B. Williams, F. J. Taylor, *Electronic filter design handbook*. New York: McGraw-Hill, 1995. 3rd ed.

[19] "Advantages of digital filters," [Online document], Available in http://www.dsptutor.freeuk.com/dfilt2.htm (current May 30, 2001).

[20] "Complete Dual-Band Quadrature Transmitters," [Online document], Available in http://pdfserv.maxim-ic.com/arpdf/2163.pdf.

[21] "Embedded Microelectronic Systems Chapter 7 – Signal Processing," [Online document], Available in http://www.ami.bolton.ac.uk/courseware/emsys/ch7/emsys07notes.html (current May 30, 2001).

[22] "Phase Locked Loop Fundamentals," [Online document], Available in http://www.minicircuits.com/appnote/vco15-10.htm (current Sept. 1999).

[23] "Super PLL Application Guide," [Online document], Available in http://www.fujitsumicro.com/pdf/PLLapp.pdf?sec=suppll (current Sept. 2000).

[24] "Synthesizer Design," [Online document], Available in http://www.minicircuits.com/appnote/vco15-9.htm (current Sept. 1999).

[25] Third Generation Partnership Project (3GPP), "UE Radio Transmission and Reception (FDD)," Technical Specification 25.101 Vol. 3.2.2, 2000.

[26] "Understanding VCO Concepts," [Online document], Available in http://www.minicircuits.com/appnote/vco15-4.htm.

[27] "W-CDMA LNA + Mixers," [Online document], Available in http://dbserv.maxim-ic.com/quick_view2.cfm?pdf_num=2387.

Acknowledgement

We would like to give special thanks to the following extremely helpful people:

Dave Devries, Manager, Strategic Applications Engineering, Maxim Corporation
Warren Hauff, Strategic Applications Engineer, Maxim Corporation
Gary Huang, Strategic Applications Engineer, Maxim Corporation
Bob Kelly, Senior Corporate Applications Engineer, Maxim Corporation
Daniel Kong, Strategic Customer Applications Engineer, Maxim Corporation
Kamal Roumani, fellow UC Riverside student who recovered damaged files
Leo Valdez, Member of the Technical Staff, Maxim Corporation
Zhongmin Yu, Strategic Applications Engineer, Maxim Corporation

We greatly appreciate Maxim Corporation for allowing us to use their facilities and resources. We are especially thankful for the help we obtained from each and every one of the engineers, technicians, and support staff who contributed to the success of this project.

Finally, we thank our Faculty Advisor Professor Mansour Eslami who made this entire project possible.

INDEx

Here, we have included only items from Parts One and Two, with just the names of participating students and their advisors in Part Three.

A
Abbreviations, 74
ABET, 3, 5, 13
Abstract, 67, 143
Acknowledgement, 70, 76, 78
Acronyms, 74
Activity logging, 29
Actual data, 79
Adams, M., 149
Addus, N., 163, 175, 197, 220, 221
Administrative, 35, 75
Advanced product quality planning and control plan, 112
Advertising, 14, 160
    cost, 82
Advisory evaluation, 60
Aesthetics, 7, 76
Affordability, 25
AIAG, 108, 109
Alternative
    approach, 75
    method, 13
    solution, 79
Ameeri, N., 163, 165, 166, 190
Analytical work, 79
Angel, 92, 104
Annuity, 87
Archiving, 30
Arts, 124, 139
ASME citation, 33
Automotive Industry, 108
Automotive Industry Action Group (AIAG), 108, 109

B
Banking, 14, 160
Behavior under stress, 154
    arbitration, 154
    mediation, 154
    mediator, 154
Beni, G., 163, 175, 176
Bertell, C., 163, 175, 176
Bhanu, B., 163, 165, 166, 190
Big picture, 13, 61, 143
Bleier, R., 88
Block diagram, 7, 35, 37, 79
Boston Consulting Group's Growth-Share Matrix, 46
    cash cow category, 46
    dog category, 46
    question mark category, 46
    star category, 46
Bribery, 58, 59
Bryant, L., 123
Budget, 35, 37, 63
Business plan, 82, 90, 110
    additional outlines, 96
    appendix, 96
    financial statements, 95
    environmental issues, 110
    executive summary, 92
    exit strategy, 96
    implementation plan, 97
    intellectual property, 93
    key people, 92
    marginal, 91
    marketing, 94
    operating, 97
    outline, 91
    problem, 93
    quality issues, 110
    sales and distribution channels, 14, 95, 160
    solution, 93
    starting point, 91
    writing, 91
Business planning, 14, 160
Business procedures, 156

C
Chasm concept, 52, 53
Chief
    executive officer (CEO), 92
    financial officer (CFO), 92
    marketing officer (CMO), 92
    operation officer (COO), 92
    technology officer (CTO), 92
Chronological order, 28
Citation
    ASME, 30, 33
    IEEE, 30, 32
Citing
    figure, 69
    table, 69

Claims, 143
Client, 23, 58, 59, 151, 152
Communication, 152, 153
Compartmentalize, 34, 41
Compartmentalization, 35
Compound interest, 83
Conference discussion, 154
Conflict
    of interest, 58, 59, 124
    resolution, 155
Constraints
    design, 7, 63, 80
    engineering, 7, 13
Continual improvement, 114, 120
    Kaizen, 120
        change, 121
        maintenance, 121
        managing change, 121
    statistical process control (SPC), 120
Continually, 109
Continuously, 109
Contract
    senior design, 7, 8, 24
Contractual
    phrases, 153
    words, 153
Copyrights, 123 to 125, 129 to 134, 144
    copyrightable, 124, 125
    fair use doctrine, 126
    not copyrightable, 125
    ownership, 124, 125
    purpose, 124
    registration, 125, 129, 131
    term, 125
    weakness, 126
Corporate planning, 45
Cost, 63, 81
    advertising, 82
    allocation, 81
    analysis, 82
    behavior, 81
    /benefit criterion, 81
    breakdown, 37, 62
    effective, 80, 81
    estimation, 81
    fixed, 81
    goods manufactured, 81
    non-recurring (one-time), 81
    overhead, 82
    prototype, 82
    recurring, 81
    sheet, 81
Course
    catalog description, 5
    credit, 4
    general guideline, 4
    grading policy, 16
    outline, 15
Critical path, 97, 115
Customer, 54, 110
    relation, 14, 117, 121, 160

D

Data collecting, 28
Date
    of discovery, 27
    patent application, 141
Davidow, W.H., 42
Davidson, S., 81
Deliverables, 7
Deming, W.E., 108, 117
Design
    comparison, 75
    constraints, 7, 63, 80
    elements of, 7, 63, 72, 76, 80
    mid-course review, 63
    professional, 149, 152
    specifications, 61, 63, 68, 74
Deuel, H., 163, 175, 197, 220, 221
Devries, D., 163, 175, 197, 220, 221
Dietrich, D.L., 123, 139, 149
Direct labor, 82
Disclosure, 58, 59, 142
Discovery (scientific), 27
Disruptive innovations, 53
Distribution, 14, 160
Doering, K., 163, 175, 197, 220, 221
Donley, E.L.R., 123, 124, 127
Drawing, 143, 154

E

Effective communication, 152

Electronic media, 30
Elements of design, 7, 63, 72, 76, 80
Elias, S., 123
Entrepreneur, 89
    common traits, 89, 90
Entrepreneurial, 89
Environmental Protection Agency, 14, 160
Eslami, M., 20, 27, 66
Essential elements, 13, 15
Ethics, 7, 14, 58, 59, 76, 160
Evaluation, 75
Everitt, W.L., 155
Executive summary, 67, 73, 74
Expansion (future), 76
Expectations, 22

F

Fair use doctrine, 126

Fairness, 22, 58, 59
Farrell, J., 163, 175, 183
Final report, 72
    checklist, 78
    citing figure and table, 69
    cover page, 73
    executive summary, 73, 74

Index

    floppy disk, 160
    keywords, 74
    labeling figure and table, 69
    structure, 73
    table of contents, 74
Financing a start-up venture, 101
    angels, 104
    banks loans, 104
    bootstrapping, 104
    debt financing, 104
    equity financing, 102
    loans from relatives 105
    loans from suppliers and customers 105
    venture capitalists 102-103
FMEA process sequence, 118
Future study, 80

G arber, R.B., 149

Giles, D., 176
Government agencies, 14, 94, 160
Grading policy, 16
Grammar, 40
Grissom, F., 123
Growing a business, 106
    acquisitions, 106
    external, 106
    internal, 106
    mergers, 106
    organic growth, 106
Growth strategies, 45
    intensive, 46
    integrative, 46
    diversification, 46
Gupta, A., 163, 219
Gyford, S., 163, 175, 197, 220, 221

H eldey, B., 42

Higgins, Professor T.J., 15, 16, 17, 18, 19, 20, 39, 65, 189
High-tech marketing, 51
High-tech ventures, 89
Historical perspective, 67
Hitchcock, D., 123
Hoang, P., 163, 219
Honesty, 58, 59
Howell, E.B., 149
Howell, R.P., 149
Human resources, 14, 160

I EEE

    citation, 30, 32
    code of ethics, 58, 59
    standard, 30, 32
    transactions, 66
Implementation plan 97
    critical path, 97
Improvement (future), 76
Initial public offering, 106
Insurance, 156
    claims-made, 157
    discovery periods, 157
    errors and omissions, 152, 157
    exclusions, 158
    expense within the limit, 157
    indemnification, 151, 157
    insured, 158
    not-insured, 58
    policy, 156, 157, 158
    retroactive date, 157
    time line, 158
    underwriter, 157
Intellectual property, 14, 80, 93, 123, 124, 139
    category, 124, 129, 160
Interest
    compound, 83
    points, 83
International Standard Book Numbering (ISBN), 123, 135 to 138
International Standards Organization (ISO 9000), 109, 110
International Standard Serial Numbering (ISSN), 123
Interpersonal skill, 153
Investors, 14, 160
IP, 124
    advice for selecting, 127
    summary, 144
IPO, 106
Irwin, K.S., 123
ISO 9000, 109, 110

J ournal, 29, 163, 219

Juran, J.M., 108, 109

K aizen, 120, 121, 122

Kanare, H.M., 27
Kelly, B., 163, 175, 197, 220, 221
Kent, J., 163, 175, 183
Kilby, Mr. Jack St. Clair, 15, 16, 17, 18, 31
Know-how/show-how, 127
Kong, D., 163, 175, 197, 220, 221
Kotler, P., 42, 43

L abeling

    figure, 69

table, 69
Labor (direct), 82
Laboratory
    notebook, 16, 27 to 30, 124, 139, 143
        archiving, 30
    session, 28
Laws of torts, 152
Lawsuit, 152
Lechter, M., 123
Legal, 14, 160
Legal liability, 152
Lessons learned, 35, 63, 64, 80
Library of Congress
    catalog card numbering, 123
    copyright office, 123
License
    agreement, 128
    clauses, 128
Limitation of liability, 151
Logging
    activities, 29
    data, 28
    electronics, 30
    retrieved, 27
    stored, 27
Lump sum, 84

M

Maher, M.W., 81

Management, 14, 46, 47, 110, 120, 160
Manual
    operator, 112
    user's, 76
Manufacturing, 14, 160
Market
    niche, 53
    vertical, 53
Marketability, 80
Marketing, 14, 42, 160
    analysis, 82
    core concepts, 43
    definition, 43
    exchange, 44
    four P's, 48
    markets, 44
    marketers, 44
    mix, 48
    needs, 43
    outline, 47
    plan concept, 47
    products, 43
    research firms, 92
    SOWT analysis, 45
    strategic planning, 44
    structure, 44
    target market, 48
    transactions, 44
    utility, 43
    values, 43
    wants, 43
Measurement systems analysis (MSA), 117, 118
Meeting expectations, 75
Member of technical staff (MTS), 36
Memo, 31, 34
Mercury News, 89
Method of solution, 74
Mid-course review, 63
MinMax optimization, 149
Moor, G.A., 42, 52, 54
Mosaic of engineering enterprise, 13, 14, 160
Mozaffari, K., 42, 50, 89, 99, 107
MTS, 36

N

Nassirpour, F., 108

Nguyen, H., 163, 175, 176
Niche market, 53
Numeral, 70

O

Open-ended, 13

Oral presentation, 36, 79
Organization, 16, 79
Original work, 79

P

Paragraph
    indented, 67
    left, 67
Parts, 82
    list, 35, 56, 76
Past tense, 39
Patent, 14, 93, 123, 139, 160
    application, 143, 145, 146
        first person, 140
        non-obvious, 140
        novelty, 140
        steps towards, 140 to 142
    categories, 140, 144
        design, 140, 144
        plant, 140
        utility, 140
    dates, 141
    disclosure, 142
    disclosure report, 142
    life, 139
    prior Art, 139
    prosecution, 142
    provisional, 140, 147
    search, 123
    subjects, 140
    term, 139
    wacky, 123
Patent and Trade Office, 139

Index

Patentability, 80
Personnel
    administrative, 35
    assignment, 35
    non-compliance, 35
Photographs, 79
Physical plant, 14, 160
PLC, 50
Positioning, 53
    concept, 54
    process, 54
Potential failure mode and effects analysis (FMEA), 117, 118
Preliminary issues, 21
Presentation
    clarity, 61
    design review, 56
    final, 159
    oral, 36, 79
Pressman, D., 123
Price per unit, 82
Price, S., 123, 139
Problem statement, 6, 37, 40, 63, 68, 74
Procurement, 14, 160
Product life cycle (PLC), 50
    consolidation, 50
    decline, 50
    emergence, 50
    end of life, 50
    fast growth, 50
    fragmentation, 50
    maturity, 50
    slow growth, 50
    stabilization, 50
    stages, 50
Product marketing, 50
Production part approval process (PPAP), 117
Professional help, 99
    accountants, 100
    attorneys, 99
    scientific, technical, business advisory board, 100
    board of directors, 101
Professional liability for engineers, 149
Professionalism, 22
Profit, 82
    expected, 82
Project
    components, 6
    deliverable, 7
    final list, 24
    ill-defined, 25
    participant, 4
    partnered (risk response), 151
    selection, 9
    specified, 6
    title, 37
    topic, 4
    well-defined, 25
Promotion mix, 48
Proposal, 30, 34
Prototype, 3, 6
    cost, 82
    evaluation, 7
Public relation, 14, 160
Purchasing, 14, 160

Q

QS-9000 process, 113
Qualitative analysis, 6, 56
Quality, 113, 114
    continuous improvement, 111
    control, 14, 160
    control plan, 116
    design and development, 111
    document and data control, 111
    management, 108
    operator instruction, 112
    policy, 110
    process monitoring, 112
    product, 112, 113, 114
    system element, 110
    system requirements (QS-9000), 109, 110
    timing chart, 115
    total, 108
Quantitative analysis, 56, 63
Quality control, 14, 108, 160

R

Recruitment, 155
Redundant, 40
References, 32, 70, 76, 78
Reflection, 24
Reliability, 7, 14, 76, 160
Report
    final, 78
Respect, 58, 59
Results
    analytical, 80
    simulation, 80
Retrieved, 27
Risk management, 149
    analysis, 150
    concepts, 150
    indemnification, 151
    issues not included, 157, 158
    limitation of liability, 151
    magnitude, 150
    response, 151
    risk control, 151
Roumani, K., 163, 175, 183

S

Safety, 7, 14, 58, 59, 76, 160
Sale, 14, 160
    expected, 82

SBA, 89
Schedule, 35
Schematics, 79
SEC, 106
Security and Exchange Commission (SEC), 106
Sensitivity analysis, 95
Sentence
    complete, 40
    dangling, 40
Seraj, A., 163, 165, 166, 190
Servin, O., 163, 175, 197, 220, 221
Shared duties, 62, 79
Shareholders, 14, 160
Shipping and
    distribution, 14, 160
    handling, 82
Simulation results, 80
Small Business Administration (SBA), 89, 96
Social impacts, 7, 14, 76, 160
Software
    list, 76
    special, 35
SOWT, 45
Special resources, 35, 76
Specified project
    faculty, 6
    industrial, 6
    student, 6
Spell checking, 40
Spitzer, N., 123, 128, 139, 144
Statistical properties of measurement systems, 119
Steudel, H.J., 108
Stickney, C.P., 81
Storage, 81, 82
Strategic planning, 44
Student group, 5
Substance, 25
Supplier, 108, 110 to 112

T

ask organization, 16, 25, 61

Team, 8, 9, 15, 17, 21, 120
Technical, 35, 37
    approach, 37, 79
    documentation, 14, 160
    faculty supervisor, 4
    presentation, 65
    problem, 6, 61, 79
    procedure, 156
    quality, 80
    reasoning, 72
    review, 67
    writing, 39, 66
Technology adoption life cycle, 51
    early adopters (EA), 51
    early majority (EM), 51
    innovators (I), 51
    laggards (L), 51
    late majority (LM), 51, 52

    niche market, 53
    vertical market, 53
Template, 66, 70
Tense, 39, 41
Test plan, 6, 35, 37, 63, 80
Top down, 61
Torts, 152
Trade
    mark, 127, 144
    secret, 126
Trademark
    formalities, 127
    protection, 127
    weakness, 127

U

S Patent and Trademark Office, 123, 139

User's manual, 76,

V

alue proposition, 54

VC, 89
Vendor, 29
Venture capital, 89
    capitalist (VC), 89
    high-tech, 89
Vertical market, 53

W

all Street Journal, 89

WARF, 123
Wassink, K., 163, 175, 176
Waste, 110, 121
Weil, R.L., 81
Whole product
    concept, 53
    planning, 53
Wilson, M., 163, 175, 197, 220, 221
Wisconsin Alumni Research Foundation (WARF), 123
Words
    contractual, 153
    multiple meanings, 40
    redundant, 40
    right sound, 40
    wrong meaning, 40

Y

ield, 85